NAVIGATING THE FUTURE

An Ethnography of Change
in Papua New Guinea

NAVIGATING THE FUTURE

An Ethnography of Change
in Papua New Guinea

MONICA MINNEGAL
AND PETER D. DWYER

Australian
National
University

PRESS

ASIA-PACIFIC ENVIRONMENT MONOGRAPH 11

ANU PRESS

Published by ANU Press
The Australian National University
Acton ACT 2601, Australia
Email: anupress@anu.edu.au
This title is also available online at press.anu.edu.au

National Library of Australia Cataloguing-in-Publication entry

Creator: Minnegal, Monica, author.

Title: Navigating the future : an ethnography of change in Papua
            New Guinea / Monica Minnegal ; Peter D.
            Dwyer.

ISBN: 9781760461232 (paperback) 9781760461249 (ebook)

Series: Asia-pacific environment monographs ; 11

Subjects: Liquefied natural gas industry--Social aspects--Papua New Guinea.
            Ethnology--Papua New Guinea.
            Kubo (Papua New Guinean people)--Economic conditions.
            Kubo (Papua New Guinean people)--Social conditions.
            Papuans--Papua New Guinea--Attitudes.

Other Creators/Contributors:

            Dwyer, Peter D., 1937- author.

Cover design and layout by ANU Press. Cover photograph: 'Entrance to Owabi Corner' by Peter D. Dwyer, Suabi, 2014.

Contents

Acknowledgements

In the years 1986 to 2014 we lived and worked with many Kubo, Febi, Konai and Bedamuni people. They cared for us and taught us. We slept in their houses in villages and forests, and walked with them through the back swamps, the foothills and the mountains. We laughed with them on many occasions and mourned with them on others. And of course, for this is Papua New Guinea, we ate with them, sharing their food and sharing ours. It has been a privilege and we thank them all.

At different times and different places we received assistance and advice from many people. Special thanks to Robin Barclay, Florence Brunois, Tom Covington, Toksie Damaga, Hugh Davies, Henry Daso, Makeya Diwo, Robyn Dwyer, Tom Ernst, Martinus Fiagone, Philip Fitzpatrick, Lucas Foifoin, Anaïs Gèrard, Laurence Goldman, Robin Hide, Garrick Hitchcock, Bob Hoad, Jerry Jacka, Siosi Kobi, Wami Kobi, Raka Kosabo, Sandrine Lefort, Paul Luoma, Evelyn Makeya, Michael Main, Craig McConaghy, Laurie Meintjes, Mike Milne, Kodu Moboi, Francis Oofoi, Willie Samobia, Jim Savage, Kerry Snelgrove, Seda Sosoaho, Romex Sowaimo, Donsi Suwo, Gosia Suwo, Henick Taprin, Jelin Tiwi, Jackson Tosiga, Kosabo Wabi, Jack Wagamoi, Chris Warrillow, Rick Wilkinson, Stanley Umosi, Willie Yofu and Colin Young. Thanks as well to expat and national staff at the Suabi and Juha camps associated with oil and gas exploration ventures during the years 2012 to 2014, the Papua New Guinea National Research Institute for affiliation during our more recent periods of research, Georgia Kaipu for her generous help in facilitating approval of research visas, Colin Filer and two anonymous reviewers for comments on the manuscript, Beth Battrick for copy editing, the University of Melbourne for granting periods of leave, and the Australian Research Council for the award of a Discovery Grant [DP120102162].

June 2017

Caveats

Some preliminary caveats are needed. These concern spelling, names and pseudonyms, quotations and some simplifying conventions that we have adopted.

We have striven for consistency, rather than linguistic accuracy, in spelling Kubo and Febi words. We lack the knowledge to achieve the latter, Kubo and Febi people are themselves inconsistent and conventions for transcribing these languages are still evolving. For example, a 1999 translation of *The Gospel of Mark* is written in 'the Koobo (Kubo) language'. Nor have we indicated nasalisation with either an 'n' or an 'm'—Febi, for example, is often written as Fembi—because, again, we lack the language skills to be consistent. And, further, although the area is relatively small and the people number less than 1,500, there are differences in pronunciation between places where population is concentrated. For example, in the west of Kubo territory we heard the local name for Cecilia River as Boiye Hoi but in the east as Baiya Hoi.

Most of the people's names that appear in this book, including those of all exploration company field employees, are pseudonyms. Place names and 'clan' names are not pseudonyms. Both, however, need further comment. To comply with government expectations, people have increasingly adopted the practice of carrying a place name with them when they relocate. For example, in 1986–87 we lived at a small hamlet that all residents and immediate neighbours knew as Gwaimasi. It was named for the small waterfall (*si*) on the stream Gwai. The 'official' name—it is recorded on the 1979 topographical map (PNG 1:100,000 Topographic Survey, Sheet 7386 [edition 1] Series T601)—is Komagato. By 1999, most residents of Gwaimasi had relocated to Mome Hafi and, more recently, some have relocated to Dege Hafi—respectively, the junction (*hafi*) of the stream Mome with the stream Dege and the junction of the latter with the Strickland River—or further afield. However, all residents of these communities are said to be, and to outsiders say they are, residents of Komagato and, in this book, when we have used that name we do so in the same general sense.

The name Suabi is itself problematic. Officially, the name is often spelled as Soabi and on documents submitted to government authorities or granting bodies that is the spelling usually adopted by local people. The 1979 topographical map (Sheet 7385) records Soabi 1 and Soabi 2—12 km apart and so named, presumably, because people from Soabi 1 had relocated to a new site at Soabi 2—but the present day Suabi (Soabi) is at a third location.

'Clan' names are spelled in multiple ways by Kubo and Febi people. We have standardised spelling of these names but do not claim that our rendering is necessarily more appropriate than some other. In the 1980s and 1990s, the Kubo word *oobi*—a mound, or gathering, of 'men' (here, meaning 'people')—connoted one or more patrifilial lineages whose members assumed, though seldom specified, genealogical connection to a common ancestor. In some cases the link to a common ancestor was supported by myth. In our earlier writing, while acknowledging instances where the actual did not match the ideal, we glossed *oobi* [*obi*] as 'clan' (Minnegal and Dwyer 2011a: 327, n. 1, n. 2). In those years people did not use the word 'clan'. Now, however, it is the other way round. 'Clan' is used routinely with the connotation of a 'bounded' assemblage of people and *oobi* is seldom used. *Oobi* is more likely to be used in circumstances that are divorced from conforming to the imagined expectations of government or resource extraction companies.

In the following chapters we have included many quotations from documents written by government officials and local people. English is not the first language of these people and for some it is their third or fourth language. We have been careful in transcribing the quoted words but have refrained from writing *sic* (representing *sic erat scriptum*, 'thus was it written') where spelling or grammar do not conform to expectations of writing in English. In one quotation we have deleted some misplaced apostrophes. Chapters, other than 1 and 3, open with a framing vignette that concerns a discrete event. These are based on notes written in the field at the times those events occurred, but are not literal transcriptions from field notes.

We use 'West' and 'Western' as glosses for global perspectives that, nowadays, are likely to be grounded in economic rationalism and individualism, and referred to as 'neoliberalism'. We use 'modernity' to refer to processes associated with 'neoliberalism', and 'modern' to refer to 'structures' and 'material goods' that are the concomitants of those processes.

Finally, Papua New Guinea currency (PGK) is based on kina and toea as analogues of, respectively, dollars and cents. In mid-1986 the exchange rate was approximately PGK1.00 for AUD1.67. Thereafter, it fell progressively to approximately PGK1.00 for AUD0.625 in late 1998 and AUD0.488 in late 2011. While it fluctuated through the next few years it remained close to 2:1 throughout this period.

Tables

Figures

1. Introduction

Towards west of Juha Gas, there are several sacred places. The biggest and the most fearing of all is called Gamlihai. At Gamlihai a Sago Palm was right in the middle of the swamp where a man called Kesomo or known by other two names; 'Wogibi and Womogosai' came out of the sago palm and went up to Mount Gamilhai and married a young woman from there.

Out of this marriage, four (4) sons were born … [they] and their descendants have hunted, fished, made gardens, made sago and have always lived together as one big extended family. They were and are a communal clan; they shared everything found on the land with each other. There were never any land boundaries in place to separate the four Kesomo sons from each other (Taprin 2007/8: 3–6).

'The Origin of Kesomo' is an unpublished 12-page booklet compiled by Henick Taprin in 2007 and 2008. The stated 'purpose' was:

to have the origin of Kesomo Clan, its descendants and its inherited natural physical features documented so that a copy can be submitted to the LNG Fembi Land Owner Association for reference and a copy kept by Kesomo clan executives so that this document will be used for any beneficiaries out of the LNG Project (Taprin 2007/8: 2).

The Papua New Guinea Liquefied Natural Gas (PNG LNG) Project is a multi-billion-dollar enterprise that in May 2014 shipped its first cargo of gas to Japan. The project is run by a consortium of multinational companies led by ExxonMobil. It is expected to produce nearly seven million tonnes of gas each year and to do so for a period of 30 years. One third of the projected income of USD31 billion will remain in Papua New Guinea (PNG) and, eventually, a proportion of this will flow to the people on whose land the gas wells are located (Business Advantage 2014).

The project will draw gas from three previously undeveloped fields and four previously operating fields, transport it via a 700 km pipeline—400 km underwater—to Port Moresby and, after processing, ship it to overseas markets. The most distant of these fields is Juha (Fig. 1.1). Five wells have been sunk there, but they are on hold and not scheduled to produce until 2020. Nor, as yet, have they been connected by a pipeline to the rest of the system. In the meantime the people who live, or lived, close to the well heads wait in expectation of huge windfalls in the form of royalty payments. Most of these people are Febi-speakers. To their south live the closely related Kubo.[1] To their north are Bogaia and Duna people, and to the northeast, across unpopulated and rugged mountains, are Huli. Kubo-speakers accept Febi, or some Febi, as the legitimate owners of the land on which the Juha wells have been sunk. At the same time, however, many Kubo-speakers seek to realign as Febi. Huli-speakers assert that, at the least, they too are legitimate owners of that land: they have a long history, and much experience, of strategising in this way (Goldman 2007).

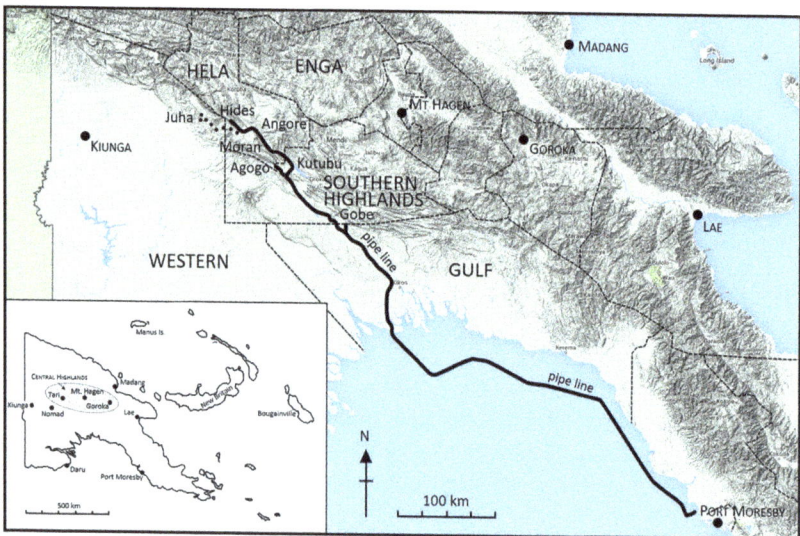

Figure 1.1: Location of gas fields and pipeline associated with the Papua New Guinea Liquefied Natural Gas Project.

Sources: Based on images available through Google Maps and Coffey Natural Systems (2009: 1.2, Figure 1.1).

---

1   In 2014, the Febi population was about 400–500, and the Kubo population about 1,000.

The cover of Henick's booklet features the ancestor Kesomo, his wife and the sago palm from which he emerged (Fig. 1.2). The names of mountains, rivers and creeks, waterfalls, sacred sites, lakes, swamps, special trees and caves on the land associated with each subclan are listed and mapped. The living and deceased descendants of Kesomo's four sons are named. They span from four to six generations.

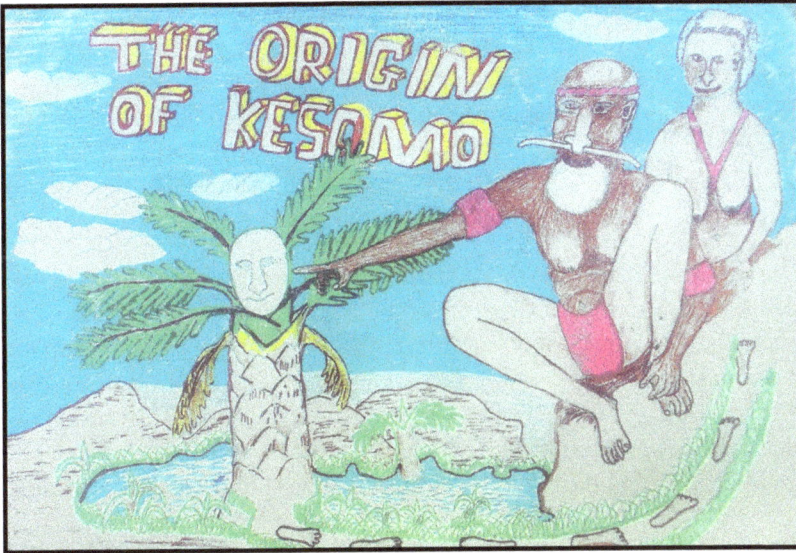

Figure 1.2: The cover of 'The Origin of Kesomo'.
Source: Artwork by Henick Taprin 2007/8. Reproduced with permission.

There are, however, complications. Some of the people named as second and third generation descendants were alive at the time the booklet was compiled. And, further, in the years 1986 to 1999, many of those named identified themselves as members of quite different clans. In some cases, husbands and wives are both listed as descendants in the same line. Henick himself, his siblings and his father, are named as descendants of two of Kesomo's sons but they hail from Oksapmin, more than 100 km northwest of the land attributed to Kesomo. By birthright they are neither Kubo nor Febi and, indeed, it is only Henick who has any direct connection to either of these language groups. In the mid-2000s Henick came to Suabi—a mission station on Kubo land—as the community health worker. He married a Kesomo woman, left with her and their son for a year or two, and returned, unemployed, in the later 2000s.

These complications might, of course, merely reflect the fact that, in PNG, asserted connections between people and asserted affiliations with land are remarkably fluid. That is, the complications may concern us, as analysts striving for order and consistency, but may give little concern to the people themselves. Those complications are, however, expressed in another way. The affiliations of people with land, with clan and subclan, and with each other that are represented in Henick's 2007/08 booklet were not those that we had imagined to be the case ten years earlier and nor were they those that we elicited six years later. Throughout this time, Kubo and Febi people were reimagining their social world, doing so with great rapidity and in two ways. They were reimagining the social 'things' of their world and the ways in which these were to be ordered. And they were reimagining the ways in which they might know those things. The ontological and epistemological foundations of their lives were changing (compare Naveh and Bird-David 2014).

This book will explore those changes. At one level, we describe the changes themselves, tracing shifts in the ways people relate to the land, to each other and to outsiders, and the histories of engagement that frame those changes. In addition, however, we are concerned with the processes through which these changes have emerged, as people seek to imagine— and work to bring about—a radically different future for themselves while simultaneously reimagining their own past in ways that validate this work.

The context in which change is currently occurring is the prospect of PNG LNG. Geo-surveys, seismic surveys and drilling—with oil and gas as targets—have been underway on Febi land for three decades. The airstrip at Suabi, opened in 1984, has facilitated these activities but, to the chagrin of Kubo people, their own land has not been focal to either exploration or discovery. These activities are driven by a variety of international petroleum companies and are regulated by the nation state of PNG. But both the companies involved and the edicts that flow from government change frequently. To secure their hoped-for future—to access the monetary benefits that one day will flow from productive wells—Kubo and Febi people must navigate this ever-fluctuating landscape. Simultaneously, however, they must find ways—both processual and structural—to accommodate being Kubo or being Febi to a future that lacks tangible guide posts. They must demonstrate to themselves and to an outside world that the past they articulate—their history of engagement with land and with each other—fits them for a future that they do not yet understand. Here too, therefore, the task they confront is akin to navigation. Out of a vast array of always intersecting esoteric and concrete memories they must discover and foreground those that fit the moment and establish a path to satisfying their desires.

The Setting

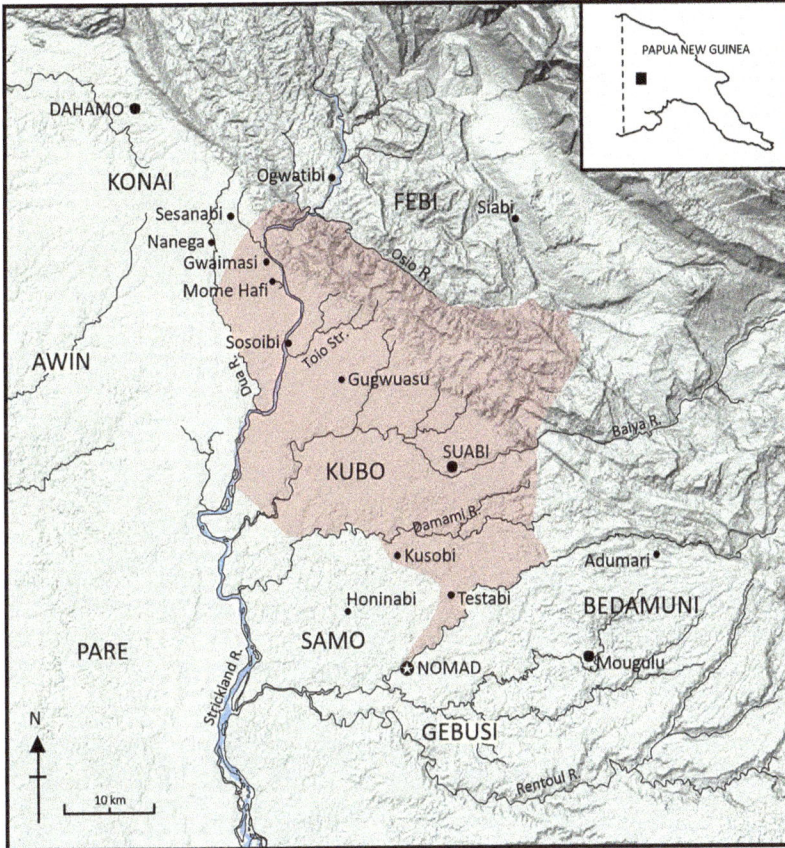

Figure 1.3: Map showing approximate location of Kubo territory.

Source: Based on images available through Google Earth.

Notes: The map shows Kubo territory (highlighted), neighbouring language groups (large font, capitalised) and primary locations mentioned in this and later chapters. Sesanabi was no longer in existence by 1986 but was re-established by 1999; Gugwuasu was established in 1987 and Mome Hafi in 1998–99. Kubo people assert that their territory extends to the government station at Nomad. We suspect that Samo, Gebusi and Bedamuni people would make the same claim.[2]

---

2   The distinct border to Kubo territory shown in Figure 1.3 is based on information elicited in 2014. Willie Samobia, of Suabi, provided the most explicit details. His interpretation was reinforced and, in places, slightly modified by others. Our prior knowledge of Kubo-Konai connections provided a more satisfactory representation in the northwest corner of Kubo territory than did information from Suabi residents. Willie Yofu—a Samo man based at Honinabi—confirmed some details concerning the border of Kubo and Samo lands. In the years to 1999 people spoke of the mix of languages represented at different communities, but at no time suggested a definitive border for any one of those language groups.

The people who we know as Kubo are the focal actors in this book. They live in a small area of the interior lowlands and foothills of the Western Province of PNG (Fig. 1.3). We first met them in January 1986 and, through the next 13 years, lived with them on five occasions for periods of between one and 15 months. Our base throughout those years was a small hamlet—initially near a waterfall on the stream Gwai, later at the mouth of the stream Mome—on the west bank of the Strickland River. At first, our research focused on the always interconnected play of social and ecological dimensions of people's lives (Dwyer and Minnegal 1992a). Later, however, the emphasis shifted to change and to the subtle processes that underlay change (Dwyer and Minnegal 2007; Minnegal and Dwyer 1999).

In 2011, after a 12-year hiatus, we returned to Kubo land. The overt changes were remarkable. The population had doubled. There were children everywhere. The survival of infants, first-time mothers and the elderly had greatly increased. Many people were now fluent in both Tok Pisin and English where, before, few had much facility with languages other than Kubo and those of their immediate neighbours. And people had travelled, and were travelling, to places throughout, and occasionally beyond, mainland PNG. On these later visits—a total of seven-and-a-half months in 2011, 2012–13, mid-2013 and 2013–14—we were based at the Suabi Mission Station. Advancing decrepitude placed our former research site out of reach; we were no longer physically equipped for, or attracted by, a two-day walk through the swamp lands that separated the airstrip at Suabi from our friends—many known since the time of their birth—who still lived close to the Strickland River. This proved hurtful to them. They felt we had abandoned them. But our research interests had shifted, and we were beholden to the grant that supported our work. Our focus was still with change. The pace of change in this area, however, and the imperative to change, had increased dramatically. The PNG LNG Project was now underway to the immediate north of Suabi and the airstrip located there served as a base from which companies accessed the Juha area or undertook exploratory work in the vicinity. The activities of these companies and, often, the physical presence of their representatives were the source of major impacts on people's lives. They suggested new understandings of the world and facilitated access to the world beyond Kubo land. They were the immediate source of expectations, desires and frustrations. They reshaped the ontological and epistemological underpinnings of Kubo lifeways. How this happened and what it means for both Kubo and Febi have been the foci of our recent research.

Changes analogous to those we describe in the following chapters have happened elsewhere in PNG. Much has been written about the ways 'desires, subjectivities and histories are being unevenly reconfigured' (Bell 2016) on resource frontiers across the nation, as land is cleared through logging or for oil palm plantations, and mines tear apart ecological and social landscapes (Kirsch 2006; Curry and Koczberski 2009; Lattas 2011; Bryan and Shearman 2015; Gabriel and Wood 2015; Jacka 2015). But the communities that are focal to those studies have, in most cases, had much longer histories of engagement with the forces of modernity than Kubo and Febi people. Indeed, it was not until the early 1960s that the colonial government established a foothold in the region where these people live. For more than a decade thereafter the focus was to the east where a larger, and seemingly more aggressive, population of Bedamuni-speakers captured most of their attention. Kubo and Febi were a backwater, occupying a thinly populated region of backswamps, forest and mountains with, it seemed, little to offer the outside world and, for themselves, a seemingly unpromising future. In places like the Highlands or Bougainville or Gulf Province, colonial administration, Christian missions, market economies, and migration for wage labour or education have been part of local lives for decades before multinational companies arrived in pursuit of resources to extract from their lands (May and Spriggs 1990; West 2006; Bainton 2010; Bacalzo et al. 2014). Most studies, too, have focused on impacts once industrial extraction has commenced. At Suabi, however, the uneven development that too often accompanies encounters with the forces of modernity remains nascent. Even by 2014 there were no mines or gas wells on Kubo land, though their territory was encircled by exploratory ventures. Nor, by this time, were there any producing mines or wells on Febi land. Kubo wait for discoveries on their own land. Febi wait for existing wells to start producing. For us, this has provided the opportunity to see the earliest phases of change, and thus capture something of the processes entailed.

Becoming Modern

In the next three chapters of this book, we describe shifts in understandings and practice over three decades, as Kubo and Febi people have sought to reposition themselves in relation to a world that, increasingly, intruded into their lives from beyond previously experienced horizons. **Chapter 2** traces the story of one community on the land of Kubo people between 1986 and 1999, and the ever-changing relationships that gave it form. Through those years, the lives of people at Gwaimasi retained much of the rhythm of earlier times. Little had changed in the material conditions of life. But people did not ignore the accumulating signs that new ways of being—previously unimagined possibilities—were emerging in the world. As the anxieties and desires these signs evoked grew, they sought ways to draw the outside world into their own and, thereby, gain access to and control over what they perceived as its future possibilities. We describe how, while material manifestations of the outside world remained minimal, people began to reconfigure previous understandings of exchange, gender relations and rights to land.

In **Chapter 3**, we turn to the histories of engagement that framed the signs of change people at Gwaimasi perceived and the responses they devised. The social landscape of Kubo has never been static; change itself was not new. But through the past century, and increasingly in recent years, new kinds of beings—with extraordinary possessions and extraordinary powers—began to appear. And what it meant to be Kubo changed as a result. We trace stories told by these outsiders, explorers and colonial officers, as they encountered and constructed Kubo as a distinct category of people. And we trace, too, the stories told by the people who now came to have that name; stories of relationships and interactions—with each other, with the land and with outsiders—that shaped the lives of individuals and groups in the present. We recount the histories of more recent times as missionaries and others came to stay, bringing with them new understandings, new possibilities and new expectations. Finally, we turn to people's encounters with bureaucracy in the guise of the PNG Department of Petroleum and Energy and other national and provincial departments that, though distant and anonymous, exert little-understood but determining control over people's lives and desires in the earliest decades of the twenty-first century.

Chapter 4 brings our story forward to 2014, depicting patterns and institutions of life at Suabi, a village where many of the historical threads traced in the previous chapter have been woven together around an airstrip, mission station and base camp for companies seeking oil and gas in the mountains to the north. While the changes we observed in earlier years emerged in response to rumours of a very different future that could be only imagined, those imaginings have taken more concrete form at Suabi. Aggregation as a larger and more sedentary community, emerging constraints on access to land, changes to marriage practices and expectations, paid employment and its effects on both access to money and the distribution of subsistence tasks among men and women have dramatically altered the rhythms of life here. New institutions—markets, schools and churches—have changed the ways relationships are negotiated and life trajectories shaped. But Suabi is more than just a village; it is a portal through which information, wealth, people and hopes flow both from and out into a wider world.

What emerges through these chapters is the increasing entanglement of local people with forces of modernity and globalisation. Two themes recur. First, people here have been left in little doubt that the future for themselves and their children will be very different from their past. This is something they welcome. They are aware too, however, that past experience provides little guidance to what will come, or to how they themselves might shape that future. Secondly, the relational imperatives that informed and patterned people's lives have eroded, with the infiltration of categorical ways of knowing the world. The money that seems key to the desired future has introduced its own logic, where the value of things—and of people—no longer lies in the relationships they mediate but, rather, in the attributes they display and the roles they play.

An orientation towards the future is not new for Kubo. They did not see events of the past as constraining, much less determining the future. Rather, the obligations and expectations that entangled a person's life were shaped not by those who came before, but by the relationships negotiated for oneself: lands traversed, trees planted, gardens made, marriages contracted, children borne and youths initiated. But the kinds of relationships possible, the range and distribution of those with whom these might be negotiated, have expanded enormously in the past three decades. How are these new terrains—their potential unknown and their nature uncertain—to be navigated?

The people we describe now see themselves on a global stage, objectified as 'Kubo' or 'Febi' through the imagined, and indeed often real, gaze of outsiders. But their position on that stage is at the margins. And, as different modes of engagement with that wider world emerge, attention is increasingly drawn to differences between people in their potential to draw on these modes. Before, all were hunters and gardeners, spouses and parents; difference lay only in the choices people made with respect to the places and other people with whom they negotiated those relationships. Now, however, only some people find paid employment, only some attend high school, only some will be recognised as potential beneficiaries of the wealth to flow from gas extraction. Which category a person falls into, which position he or she holds, is in many ways determined by others; companies employ workers, churches designate pastors and the government appoints community health workers and teachers, while village councillors are elected at the whim of a local majority. The attributes of persons thus become more salient, in deciding what can be expected of them, than the relationships they have entered into. As a consequence, what has been done in the past matters in new ways; it becomes objectified, open to evaluation.

A sense of rupture between past and future, and of differentiation in the present, has been recognised by others as key to a modern sensibility (Englund and Leach 2000; Knauft 2002a). People at Suabi know that the future will be different. But how different will it be, and in what ways will it differ from the past and present? They are intensely engaged in imagining what that future might be like, evaluating different possibilities and working to bring the future they desire into being. In doing so, the past itself becomes a point of reference—to be rejected, perhaps, or restored, but always as something against which new possibilities are evaluated. In the next two chapters of this book, we trace ways in which people at Suabi in recent years have been navigating the uncertainties that arose from encompassment within, and their own attempts to encompass, the promises and threats of a modern world.

Navigating Change

Kubo and Febi people have always lived in a world of uncertainty. They seek to recognise and act on the ever-changing opportunities that emerge in the world. Always, however, this is done in the knowledge that outcomes may not be as hoped; other agents, too—human and non-human—are pursuing their own agendas, and this may alter the consequences of action. But there are new agents to be dealt with now—church, state, company—whose decisions have profound effects on both available opportunities and what must be done to seize those opportunities. The world that people must navigate 'in relation to the push and pulls, influence and imperatives, of social forces' (Vigh 2009: 432) is now more fluid. Like the West Africans with whom Henrik Vigh worked, people at Suabi:

> spend a great deal of time debating how global, regional and local influences and conflicts will affect their lives, what spaces of possibility will emerge or disappear, what trajectories will become possible and what hopes and goals can be envisioned (ibid.: 422).

Vigh's concept of 'navigation'—of moving through an environment that is itself always in motion—captures well the tactics and strategies deployed by Kubo and Febi people as, in recent years, they have felt their way through 'the immediate convulsions of a fluid environment whilst simultaneously trying to gain an overview and make [their] way toward a point in or beyond the horizon' (ibid.: 429). Their actions are concerned with 'both the socially *immediate* and the socially *imagined*' (ibid.: 425). But, simultaneously, their tactics have changed the environment for future action, reconfiguring the social world in key ways.

In seeking paths to a future that was itself barely imagined, a future that held the promise of great wealth and potential influence, people at Suabi had to navigate the complexities of government and company bureaucracy—complexities that seemed impossible to pin down, were ever-changing, and could be known only through constant attention to the rumours that trickled in from outside. To establish their credentials as legitimate actors entitled to negotiate relationships and enter into agreements they drew up lists of named people, planning to register Incorporated Land Groups that were eligible to receive royalty payments. In this process they reworked the organisation of households and the relationships between men, women and land. They drew maps, and compiled catalogues of mountains and streams and swamps that belonged to groups of people that they now represented as bounded in ways that had never obtained in the past. They drew upon the discourses of the government and corporate sectors: tribes, clans and subclans; councils and committees; companies, chairmen, treasurers and agents. In all these ways they explored different ways of being in the world, with the outcomes that they brought very different social entities into being and, ultimately, themselves became different kinds of people. It is these changes, and the navigation of imagined futures that framed them, that are the themes of **Chapter 5**.

Although people drew on the language of outsiders to pursue their objectives, the logic that informed their efforts was not that of either the state or resource companies. The relationality that grounded identity in the past continued to inform decisions. The process of bringing new forms of collectivity into being has entailed revisiting—and re-visioning—past actions and interactions. **Chapter 6**, therefore, focuses on the ways in which people at Suabi sought to validate action in the present by selectively drawing on the past—by navigating a past that could be always read in other ways. The readings foregrounded in 2011–14, however, projected new interests and imperatives into that past. As people sought to position themselves in relation to opportunities offered by global capital flows, they mobilised connections established long before. They will have always done this. But, in drawing on a past that now played out in a very different world, they were rereading events and understandings that had once shaped that past. They gave overt expression to this in both the resolution of disputes through the new mechanism of village courts and in reshaping mythological pasts as they reshaped social groups.

In the former domain, at public court cases concerning unreciprocated marriages, inappropriate liaisons, behaviour that risked angering spirits, or accusations of sorcery, protagonists appealed to events that had occurred decades before or drew on past practices, sometimes long abandoned, to critique actions in the present. By these selective appeals to the past, they sought to mobilise compensation payments in the present. The effect was to redistribute money that entered the community along paths that not all could directly access.

In the latter domain, people devised symbols that were intended to ground and sustain the new social collectivities that they were bringing into existence. They did so in the form of logos, using elements drawn from the mythological past. Mythologies that previously had woven people together in an unbounded relational field were now, themselves, being deconstructed. Different groups of people began to creatively define the limits of the cosmological by declaring selected stories or selected constructions of stories to be properly and exclusively their own. They were, in effect, corporatising social *and* cultural identity (compare Comaroff and Comaroff 2009).

In both domains, then, Kubo and Febi people drew selectively from their remembered pasts, both secular and sacred—often foregrounding some details while suppressing others—to fit their lives to an ever-changing present. For them, the present is deeply imbued with desires that are oriented towards a future imagined in terms of 'development'—a future in which they are no longer 'remote' and forgotten, and in which wealth that is perceived as 'rightfully' theirs is given material expression. In 2011–14, that imagining was closely bound up with the presence at Suabi of a relatively large campsite that served as the base for exploratory geo-surveys and seismic surveys being undertaken by several different petroleum companies.[3]

Encountering Otherness

The concluding chapters of this book turn to ways in which people at Suabi framed their encounter with 'otherness' in the form of the exploration camp, and with how this informed a deeper refiguring of local ontologies and epistemologies. We develop more explicitly conceptual threads that emerged in earlier chapters. Most importantly, we emphasise the contingency, plurality and interrelatedness of ontologies and epistemologies. Drawing on Heidegger, Tim Ingold (2010: 4) writes of a 'thing' as having 'the character not of an externally bounded entity, set over and against the world, but of a knot whose constituent threads, far from being contained within it, trail beyond, only to become caught with other threads in other knots'. An 'object', by contrast, 'stands before us as a *fait accompli*, presenting its congealed, outer surfaces to our inspection. It is defined by its very "overagainstness" in relation to the setting in which it is placed'. The emphases here are, respectively, on what we distinguish as the relational and the categorical. But where, it seems, Ingold favours a view that there are no objects in the world, there are only things, we favour the view that at different times and places, and

---

3   People at Suabi often referred to the mix of companies that were based there as 'Company' and the campsite as 'camp', 'Company camp' or, occasionally, 'mining camp'. In this book, in contexts where particular company identities are not relevant, we follow the local practice of using 'Company' to denote the collective.

irrespective of the possibility that Ingold is correct, people may come to know things as objects. This was evident in the ways that people at Suabi related to, and made sense of, the exploration camp and the company representatives based there.

People at Suabi had high hopes for future benefits from the PNG LNG Project, in the form of royalty payments and business development grants. In their understanding, those benefits would be provided, either directly or indirectly, by 'Company'. And, for them, Company had a very material, and personalised, presence (compare Golub 2014); a presence that offered opportunities in the present for those who were able to discern and act on them. The camp at the Suabi airstrip was woven into everyday life. These everyday interactions, and the access to benefits that they facilitated or constrained, are the focus of **Chapter 7**. They reveal much about how local people understood the nature of Company and their relationship with it. But these interactions were framed, too, by the understandings of Company representatives, and these understandings did not always coincide with those of local people. Nor were they constant. On the one hand, it often seemed that there was an air of arrogance in the performance of Company representatives, a sense of presumed superiority that precluded any need to either pay attention to local concerns or communicate their own intentions. On the other hand, it often seemed that local people were complicit in the dictates and ideological persuasions of Company—working to project an image of themselves as 'cooperative', welcoming, non-demanding and non-threatening hosts. To us, as observers, these perceptions were discomforting. But they did not do justice to either party. In this chapter, we show how the apparent 'arrogance' of Company and the apparent 'complicity' of local people were mutually imbricated, each more a product of the 'friction' between two ontological systems than an expression of either system alone (compare Tsing 2004).

In **Chapter 8**, we shift scale, to explore how not just interactions with those at the exploration camp, but also the ways in which the structure and operation of that camp affected local ontologies. Through earlier chapters we trace shifts in the ways that people at Suabi interacted and sought to manipulate relationships, not only with each other but also with the new beings, things and powers that have appeared in their lives through the past three decades. But lurking behind these changes there have been deeper shifts in modes of understanding. In this final chapter, we pull together some of the threads woven through previous stories, to highlight the interplay of ontologies and epistemologies in the strategies that people deployed to navigate both future and past, and to manoeuvre in the present.

Our perspective on both ontology and epistemology—on the known and the knowing (Dewey and Bentley 1975)—is that neither is static and unchanging, and that each arises and may alter in the course of living in the world. Our perspective is, thus, phenomenological (compare Bird-David 1999; Ingold 2000; Blaser 2009a, 2013). It is not, we consider, in accord with the recent and influential perspectives that have been developed by Phillipe Descola and Eduardo Viveiros de Castro.[4] The 'things' in the worlds of people are not pre-given. They emerge and consolidate or dissolve as people engage with that world. Nor are the ways in which those things are ordered pre-given. As Gregory Bateson wrote: 'The division of the perceived universe into parts and wholes is convenient and may be necessary, but no necessity determines how it shall be done'

---

4   Discussion of the recent 'ontological turn' in anthropology is concerned, especially, with the substantive and challenging contributions of Phillipe Descola (2006, 2013) and Eduardo Viveiros de Castro (1998, 2004a, 2004b; see Carrithers et al. 2010). Descola offers a structural analysis under which identity and difference on two intersecting axes of 'physicality' and 'interiority' generate a fourfold typology of ideal ontologies; models with 'no empirical existence' (Taylor 2013: 201). His concern is with ways in which 'things' may be ordered, more than with the 'things' per se. His axes of 'physicality' and 'interiority' resonate, perhaps uncomfortably, with an understanding of body and mind as distinct categories and he acknowledged this in writing that 'according to developmental psychology, the awareness of this duality is probably innate and specific to the human species' (Descola 2006: 147). Descola's ideal ontologies do not arise in contexts of engagement with environment— that is, in practice—but are given, or made possible, by pre-existing mental templates. Viveiros de Castro, working initially from Amazonian ethnography, offers 'perspectivism' as an ontological type under which humans and animals see themselves, other beings and the 'things' they and other beings are afforded by environment in precisely reciprocal ways; thus, for example, a jaguar sees itself as human, sees humans as animals, and sees the blood it drinks as humans see manioc beer (1998: 470). To Amazonian peoples, he wrote, 'the original common condition of both humans and animals is not animality but, rather, humanity' (2004a: 465). Viveiros de Castro enlarges on this perspective in proposing 'multinaturalism' as a more general, and speculative, ontology under which all the inhabitants of the world share the same 'culture' but have many different 'natures'. These are challenging ideas. We agree that different people may apprehend the 'things' of the world in different ways and that this may result in a 'communicative disjuncture where the interlocutors are not talking about the same thing, and do not know this' (Viveiros de Castro 2004b: 8). But we have one primary concern. Not all beings or objects are understood to fall within the ambit of perspectivism and, indeed, there is variation across Amazonia with respect to both those that do and the human persons who hold the knowledge of their status (Viveiros de Castro 2012: 48, 53–4). Beings with great symbolic or practical importance receive most attention, and context may thus inform observed variation. What, therefore, is the ontological status to Amazonians of those beings or objects to which perspectivism does not apply? To them, is ontology itself to be understood as multifaceted? We accept that these authors, Viveiros de Castro in particular, provide access to a critical and radical anthropology that acknowledges that we all potentially live in 'multiple realities' and can be other than we are (Hage 2012), but their take on, at least, ontology is not that which informs discussion in this book.

(1979: 42). The 'ontological bricolage' (Gewertz and Errington 2016) that results must be understood as an expression of 'shifting, situational, melded, often self-conscious, and sometimes critically appraised ways of being and acting in the world' as people 'slide in and out of multiple frames of reference' (ibid.: 375–6).

One way in which people give substance to the 'things' of their world is by naming them. We open this chapter, therefore, by discussing ways in which Kubo and Febi people have altered ways in which the bestowal of names reveals how individual people and groups of people perceive themselves and are perceived by others. As in so many other domains of their lives, the changes reflect erosion of a pre-existing and predominant relational ontology and epistemology and an infiltration of a categorical ontology and epistemology. And, always, these changes, though never absolute, arise in practice. They arise, are tested and rejected or affirmed, in the course of everyday experience—in the course of people's everyday engagement with the world.

It has been in the course of their engagement with representatives of petroleum companies that the people who live at Suabi have had their most immediate and direct experience of other ways of living and other ways of knowing the world. They themselves, as hosts and workers, have contributed to the sustainability of the resource extraction ventures that have operated on or near their lands. Their efforts have flowed outwards to a globalised world, to a distant world that has no sense of the efforts that people in a remote, lowland forested corner of the world contribute on its behalf. In return, the ideological persuasions of that world, the assumptions of science and the market place that give it certainty—in short, the ontological and epistemological foundations of that world—have penetrated into the very being of the people who live at Suabi. Those persuasions and assumptions underwrite the structure and performance of the exploration camp at Suabi. They pattern the behaviour of the men who work there. To local people they provide models of how the world could be and of what might be done, or should be done, to achieve desired ends. These processes, and their consequences for people at Suabi, are the central concerns of this chapter. We draw on the diverse understandings of, among others, Anna Tsing (2004, 2009), John and Jean Comaroff (2009) and Mario Blaser (2009a, 2009b) with respect to ways in which the hospitality or products of local people may service the wants of distant elites, the ways in which the ideological persuasions of intruding outsiders may infiltrate the understandings of those whom they intrude upon, and the ways in which the meeting of diverse ontologies and epistemologies plays out on a political stage where, ultimately, it is people's everyday experiences that shape their world-view.

The processes we write about tell of ways in which the people at Suabi are comprehending their own social arrangements in new ways and, indeed, how they are comprehending themselves in new ways. They tell, also, however, that there is no absolute rupture. Issues of identity and concomitant issues of practice run deep. At Suabi, as so much has changed, people continued to engage with an environment that, on the one hand, they understood to be unpredictable and, on the other, to 'give' without a requirement that they reciprocate (compare Bird-David 1990). Nor did people ever forego their deep concern that they might be ensorcelled by others within their community, or themselves be accused of sorcery by those others. Though they were reaching out to a world beyond the local, and though the ontological and epistemological framing of their world was in flux, they remained committed to expressions of relationality that, for a time at least, seemed to obviate that concern.

The changes we describe in this book, whether surface or deep, were not forced upon Kubo and Febi. They learned that life was different elsewhere. Outsiders came to their land and, in their own ways of living, both revealed something of those differences and enhanced desire for what might be offered by embracing such alternatives. The changes that followed were, in large part, an outcome of the actions of the people themselves. A new world is emerging for Kubo and Febi people, a world that they themselves are building, a world in which they are emerging as new kinds of subjects.

2. Gwaimasi: 1986–99

The woman and her children came out of the forest to the edge of the river. That is where they saw the fish. It was dead and had washed onto the bank. It was longer than a man's arm and without scales. The body was reminiscent of an eel but the head was peculiar; it was large and the mouth was underneath. The woman thought that the dead fish was an eel with a man's head, and was frightened. She hurried back to the village to tell others what she had found. Then she led some men to the place. They looked at the dead fish but did not touch it. None of them had seen anything like it before. But it reminded them of a story they had heard when they were young—the story of Kongwa, a huge eel that could take the form of a man and which lived in the river and the largest tributaries. Sometimes Kongwa threatened people who were crossing the river by dugout canoe and at other times, when the water was rough, he swam beneath the canoe and guided them to safety. The men concluded that this fish was a child of Kongwa. They decided it had died because a mining company, located in mountains hundreds of kilometres to the northeast, had poisoned the river by adding chemicals to the water. Often, through the past few years, people had caught catfish with ulcerated sores on their bodies and these too had been attributed to pollution. The dead child of Kongwa was understood to be a sign that people should no longer eat fish from the river. It was a sign that the world was changing in ways over which they had no control. But it also reminded the men of rumours that had reached them from a variety of sources; rumours which suggested that, elsewhere, there were other people who were benefiting from the changes. The dead fish created both anxiety and desire.

The men used sticks to push the body of the fish away from the bank of the river and into the current. It was caught in an eddy, held there briefly, and then swept downstream. No one found another fish of the same kind.[1]

When we first arrived among Kubo people, in 1986, their lives retained much of the rhythm of earlier times. Metal was replacing stone, and cloth replacing string skirts, but material and social life seemed little changed. Even the new tools and clothes arrived along old pathways. The Kubo we came to know were experimenters, responding to the always unpredictable opportunities that the forest offered, playing with alternative ways to build an oven or ornament a dance costume. Relationships negotiated in the past did not constrain who one should marry, or where one should live; it was always possible to seek out new ties, to construct new identities. But modes of relationship, the expectations and obligations that accompanied them and the actions that produced them, were familiar to all. Through the next 13 years, however, signs began to accumulate that new ways of being—previously unimagined possibilities—were emerging in the world.

In this chapter, we trace the story of one Kubo community through those years, and the ever-changing relationships that gave it form. Though little changed in the material conditions of life at Gwaimasi between 1986 and 1999, the anxieties and desires evoked by signs like that dead fish found on a river-bank in 1997 gradually transformed the ways people made sense of their lives.

---

1   A likely candidate for the dead fish found in the Strickland River is an immature river shark, *Glyphis garricki*, which is known from northern Australia and Papua New Guinea (PNG), and which had ranged far to the north in response to the devastating drought of 1997 (Minnegal and Dwyer 2000a; Compagno et al. 2008; Thompson 2011).

Relationships in Place

Sinage, Monu and Tobu were initiated together in the early 1960s. They were young men of three different *oobi*—different named, patrifilial groups.[2] The venue was a longhouse near the junction of the streams named Mome and Dege. They were decorated, received ceremonial arrows from their initiation sponsors and committed to several food taboos that were to persist to the time when the first garden they themselves made after initiation had been harvested and was giving way to secondary-growth tree ferns. Uhabo, who was of a fourth *oobi*, should also have been initiated at this time, but he had eloped with the widow Umode and, in the practice of Kubo-speakers, a man who married before he had been initiated could never be initiated.

The streams Mome and Dege drain backswamps in low-lying, densely forested land immediately west of the Strickland River in the interior lowlands of the Western Province of Papua New Guinea (PNG; see Fig. 1.3). The altitude is 100 m above sea level and annual rainfall may reach 6 m. To the east and north the swamps give way to forested foothills and rugged mountains. Sago palms, nut trees, wild pigs, cassowaries, fish and crayfish are abundant. The swamplands and foothills are occupied by Kubo-speakers, the lower mountains by Febi-speakers. To the west are Konai and Awin-speakers and to the south and southeast Pare, Samo and Bedamuni-speakers.

As co-initiates Sinage, Monu and Tobu maintained a long-term association. They addressed each other as *samo*—'co-initiate'. As particular longhouses deteriorated and new ones were established in different places they moved together. Uhabo travelled with them. They married. Monu married the widow Haga. Sinage married Wanai who was one of Tobu's sisters. Uhabo's first wife, a classificatory sister of Sinage, died and he remarried to another of Tobu's sisters, Fafobia. Tobu was the

---

2   A Kubo *oobi* is, literally, a 'man mound'—an assemblage of people brought together, as a brush turkey (*djago*) rakes together the leaves on the forest floor to make a mound (*djago bi*) in which to incubate its eggs. Golub (2014), writing of Ipili in the highlands of PNG, describes analogous assemblages as 'activity sets'—a group of people who have come together for some purpose, its membership always contingent though certainly not random. Among Kubo, who one is born of may influence which *oobi* one is aligned with; children are, after all, more likely to associate with their birth family than with others, at least in their early years. But where, as was too often the case for Kubo, men die young and widows remarry, a child may well be raised by his or her mother's new husband of a different *oobi*. Where one chooses to garden, or to hunt, matters too. The places visited, and the stories learned from others, and with others, also matter. *Oobi* identity is strong, and persistent; it aligns people with others, and with places. But it is not exclusive, and it is not immutable.

last to marry. His marriage balanced Uhabo's second marriage, for he married Babio who was stepdaughter to Uhabo. At that time exchange marriages—preferably by two men who married each other's sisters—was the usual pattern among these people.[3] And indeed when, much later in 1986, Tobu's younger brother Sidine married, his wife was classificatory sister to Sinage. She was of the *oobi* into which, as a boy, Sinage's father had been adopted. The two couples formed through an exchange marriage commonly chose to live and work closely together. They were, after all, more closely related to the children of their exchange siblings than to those of other brothers or sisters.

In 1979 these men, with their wives if they were already married, were part of a group of perhaps 60 people—speakers of Kubo, Febi and Konai languages—who assembled as a new community close to the stream Sigia, on the land of Konai. This was Sesanabi and the people who came together here were responding to the call of an Evangelical Church of Papua New Guinea (ECPNG) pastor who had come overland from the town of Kiunga. An initiation was held in 1984 and soon afterwards, when the pastor had left, some deaths had occurred, and sorcery concerns were in the forefront of people's imaginings, the community fragmented and dispersed. Monu and Uhabo with their affiliates—sons to the previous husband of Monu's wife, a man of Uhabo's *oobi*—moved to a temporary garden house downstream on Sigia, close to its junction with the Strickland River. Sinage and Tobu with their affiliates—*oobi* brothers to Tobu—moved further south to a temporary house where sago palms were near at hand. They were living at these places when we first met them in January 1986. Most of those aligned with Sinage and Uhabo were of, or affiliated with, Gumososo *oobi*. Most of those aligned with Sinage and Tobu were of, or affiliated with, Gomososo *oobi*. They had decided, however, to regroup, build a longhouse at the place where the small stream Gwai tumbles over a waterfall to join the Strickland River, and reassert themselves as a community. The relational bonds that tied these families together were those of initiation and exchange marriages.[4]

---

3   In the 1980s, no bridewealth payment was involved in marriage exchanges, though the men did, for some time afterwards, make occasional gifts of meat to their wives' fathers.

4   Two other young men were initiated at the same ceremony as Sinage, Monu and Tobu. One had died by 1986. The other did not live with his co-initiates at Gwaimasi in 1987 but, rather, with his wife's people at a neighbouring Konai community. He often visited Gwaimasi through that year, however, and in later years rejoined his co-initiates. The 1984 initiation at Sesanabi was the last held in the west of Kubo territory. Initiations were held in eastern Kubo territory in 1987 and 1991 but there have been none since 1991. Several youths and young men from western *oobi* were participants in the 1987 ceremony.

Our visit in January 1986 was to introduce ourselves and ask whether we could return. We did so in August of that year and stayed for 15 months. By this time the people we had met had, as they said they would, built a longhouse at Gwaimasi on the land of Gumososo people (Fig. 2.1). Two families had also built separate houses, and other houses were under construction. People were adopting the new practice of establishing a hamlet at which traditional communal longhouses were replaced by family houses (Fig. 2.2).[5] Through this intense period of building, people relied on sago as their staple carbohydrate, though sometimes they would make the two-hour walk to harvest remaining bananas at gardens they had abandoned at Sesanabi. They had prepared and planted new gardens, however, at which the primary crop of bananas was coming into production by the time we arrived to stay.

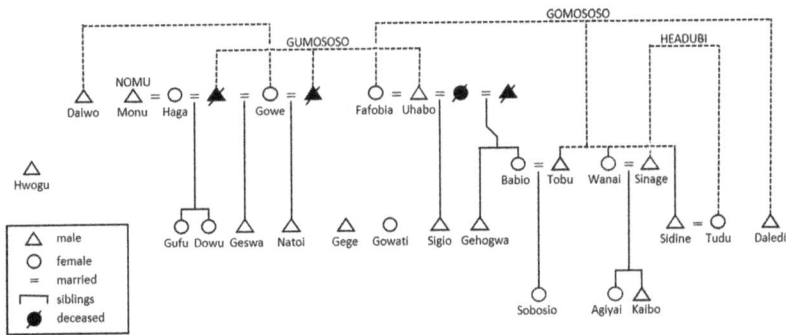

Figure 2.1: Residents of Gwaimasi at August 1986.

Source: Authors.

Note: Dashed lines show *oobi* affiliations. *Oobi* names are capitalised. Daledi was of Gomososo *oobi* but not of the same lineage as Tobu and his siblings. Gege and Gowati were orphans in the care of the widow Gowe. Changes to November 1987 were as follows: the Febi males Daiwo and Hwogu departed; Gowati departed in marriage to a man at Suabi Mission Station and, in a balancing marriage, a girl from Suabi married Geswa and came to live with him; Gehogwa departed for school in January 1987 and was absent thereafter; a young Headubi man, and later his new wife, joined the community on a temporary basis; and children were born to Haga, Wanai, Tudu and the unmarried young woman Gufu, but two of these four had died by November 1997.

---

5   The longhouse at Gwaimasi had been abandoned and demolished well before October 1991.

Figure 2.2: Gwaimasi village, 1987.
Source: Photograph by Peter D. Dwyer.

Livelihood

The people living at Gwaimasi were hunter-horticulturalists (Dwyer and Minnegal 1991, 1992a, 1994). They hunted and trapped wild pigs. They used palm-wood bows and cane-grass arrows, putting the finishing touches to the former with flaked chert tools. They built hides on the ground and in trees from where they shot cassowaries and smaller birds. They fished with hooks and line in the always silt-laden river and the clearer larger streams. In the latter, men shot fish using a length of strong wire that was propelled by rubber.[6] Occasionally they prepared poison

---

6   In 1986–87, the Kubo fish spear (*audi*) was made from modern materials. These, however, had simply replaced an earlier design that had been first described by W. Patterson and attracted the interest of senior administrators. Patterson (1969a) wrote: 'An interesting type of fish spear was noted in the upper Kubor area, the spear itself was simply a bamboo shaft with a single point although they may have multiple prongs. The back end of the shaft was attached to a larger piece of bamboo and the larger piece of bamboo was then passed over the end of the shaft. The attaching piece between the shaft and the bamboo tube was a piece of native rubber, locally grown and cured. To fire, the shaft of the spear would be drawn back through the tube and then released, the tension of the rubber driving the spear point into the body of the fish. … The rubber substance was gained from the centre of a type of [vine] which was placed in a hollow tube of bamboo and then dried or cured in the smoke over a slow fire, the result is a tough and highly elastic piece of rubber of roughly ½ inch diameter, this is then cut to the required length – approx 7 inches for a fish spear'. In a later report, where he corrected 'cane' to 'vine', Patterson noted that the spear was used by people of the upper Strickland and the Cecilia [Baiya] Rivers, that 'in some instances the curing process may be started by spreading the sap thinly over the forearms where body heat would start the curing process' and that 'a rubber

from the roots of particular plants and used this to take larger hauls of fish. Crayfish and giant shrimps were speared in their burrows. Smaller animals—snakes, lizards, frogs, some insects—were taken more casually, adding variety to an always protein-rich diet.

Nearly half the carbohydrate that people ate was flour processed from the pith of both wild and planted sago palms. Some of these palms were felled to incubate the fat-rich larvae of weevils and others were used as traps for wild pigs which could eat their way into the trunk but could not reverse out again. Breadfruit and the nuts of *Terminalia* and *Canarium* were available in season and seeds from these latter trees were regularly planted at garden sites and the clearings where longhouses or hamlets were established. Fruit from a dozen or so red and yellow varieties of fruit pandanus (*P. conoideus*), all cultivars, was available in the later and earlier months of each year. The fruit, steamed in earth ovens and squeezed to remove the seeds, yielded a rich oily sauce that was eaten with sago. Fern fronds and fungi were other valued wild vegetable foods.

On the levee banks of the river and larger streams families made banana gardens at which the ground was cleared, crops planted and trees then felled on top. It was 15 or more years before these garden sites could be reused (Dwyer and Minnegal 1993). A few smaller gardens, usually fenced against the depredations of wild pigs, were prepared for tubers—taro, yams and a little sweet potato. Sometimes yams were planted in the burned-out mounds of leaves and twigs that had been scraped together by jungle fowl to incubate their eggs (Dwyer and Minnegal 1990). Sugar cane, a variety of vegetables (for example *Hibiscus manihot, Rungia klossi, Setaria palmifolia*) and lowland pitpit (cane grass, *Saccharum edule*) were found in nearly all gardens. Some small gardens, however, were dedicated to lowland pitpit which is, perhaps, the most seasonal of the crops that Kubo people grow. *Hi a*—pitpit time— is the name given to this season and when the inflorescences of this plant are abundant, and ready to harvest, they often form the basis of small feasts. The saplings of tulip trees (*Gnetum gnemom*) are left when gardens are felled and the young leaves are gathered, often by children while their parents weed or harvest bananas, to be cooked and eaten.

---

will last only 18 months to two years before requiring replacement' (1969b; supplementary letter dated 11 August 1969). In 1987, Hami demonstrated the process to us, though he cured the rubber on his chest and by lying in the sun. We did not use the sample he made but it retained elasticity for nearly ten years. Beek (1987: 93, Fig. 26) described and illustrated a similar fishing spear used by neighbouring Bedamuni people. It used wire prongs attached to a length of cane and was propelled by 'modern' rubber. He wrote that it was 'essentially new' to these people.

During our first year at Gwaimasi, the only domestic animals kept by the people who lived there were pigs and dogs. The former, which numbered between seven and 13 through 1986–87, were sows or castrated males that were either captured as wild piglets or born to domestic sows that had been impregnated by wild boars. They were individually owned by men, women or children. Responsibility for their care fell to women, though a woman would avoid this responsibility in the later stages of pregnancy or the early months of lactation. Pigs were a threat to unborn babies and young infants, and care was taken to minimise the likelihood that a pig might feed on the urine or faeces of a young child. The bond between a domestic pig and its carer was exceptionally close. For up to 18 months the pig might accompany its carer when she went to gardens or to process sago, or be taken to the forest to forage while its carer stayed nearby. Thereafter, the animals were left to forage alone in the back swamps, though they were visited by their carer to be groomed or, if they were sows that had given birth, to capture the piglets. Domestic pigs were not tame; they were imprinted on their carer and few other people risked going near (Dwyer 1993; Dwyer and Minnegal 2005). They were killed and eaten at feasts and curing ceremonies and, occasionally, simply because they were being a nuisance. Dogs were valued as hunting animals. Four or five would accompany a man who was seeking pigs. There were usually around 20 dogs resident at Gwaimasi in 1986–87, a little less than one per person.[7]

Through 1986–87, there were five attempts to raise a wild-caught cassowary chick with thoughts of selling the birds to people in the highlands. All attempts failed: the birds died, escaped from the string bags in which they were being held, or were killed by dogs. People said that attempts to rear chickens had been similarly unsuccessful, at Gwaimasi and at neighbouring communities, due to the depredations of dogs. By 1991, however, after a distemper epidemic had killed 15 of 19 resident dogs and with more Kubo people aligned with the Seventh-day Adventist (SDA) mission, both cassowaries and chickens were being raised (Dwyer and Minnegal 1992b). In 1991, one cassowary at least 120 cm tall was in captivity at Gwaimasi, and another three had been raised to a good size in the previous few months. At this time, Gwaimasi was also home to a rooster, a young pullet, and at least six chicks ranging from a few weeks to a few months in age; all had been procured from other places and

---

7   In 1986–87, the dogs at Gwaimasi howled, rather than barked, and frequently chorused. We consider them to have been representatives of the dogs found as domestic animals at all altitudes of New Guinea, and as wild animals in high altitude locations, at the time of European colonisation (Dwyer and Minnegal 2016).

none as yet had been hatched locally. The captive cassowaries were not tamed, but held in strong pens with food carried to them daily. Chickens, by contrast, were raised in ways analogous to pigs, with a strong bond developing between the bird and its carer. But the cassowaries raised in 1991 were no longer destined for distant markets, their meat intended rather for use in local exchanges, either for sale at markets or as formal prestations. Chickens, too, were intended for consumption on special occasions; there was little interest in harvesting their eggs.

All these key resources came and went—some seasonally, others unpredictably. The environment was rich but procurement was boom and bust. Sago flour was the primary carbohydrate at some times, bananas at other times. The meat of wild pigs was not always available but fish were a reliable, and regular, source of protein. Some resources offered immediate returns, with others there was a delay. But no one was concerned by these erratic fluctuations. People took what was available and, when one potential resource was unavailable, turned to others. They were always confident that the environment would continue to give (Bird-David 1990; Dwyer and Minnegal 1998).

New gardens could be established at any time of the year; there is little seasonality here, in rainfall or temperature. Sago, too, could be processed at any time. The resultant flexibility meant that there was little patterning in activities. Individuals, and households, were likely to be engaged in very different tasks on any day—some felling gardens, others processing sago, one man hunting, another fishing, yet another visiting friends at a neighbouring village. All attended to signs of what was available, or soon would be, in the forest or at gardens—which sago palms were showing signs of flowering, where a pig had been rummaging or fruit dropped to the ground by feeding birds. They watched and listened, too, for indications of what others were planning to do in the next few days; if one man had thoughts of hunting pig, another might be better advised to turn to some other task. But these were decisions for individuals, attending to signs that the time was right to undertake a particular endeavour (Minnegal 2009). When nut trees bore prolifically, as they did every few years, then everyone might harvest these in the course of other activities. And in the months when pandanus fruit were available these too were carried to the village on most days. Only *hi a*—pitpit time—gave some consistent form to the year, a consistency that allowed this season to become a marker of passing years. As the inflorescences of this cane-grass began to appear, towards December each year, people prepared for feasts that brought neighbouring communities together. Even here, however, there was flexibility. Larger feasts—one held at Gwaimasi in 1986 (Fig. 2.3), and another at Gugwuasu in 1987 to which Gwaimasi residents contributed much smoked game—brought a common focus to activities, but families remained free to schedule tasks to suit household interests.

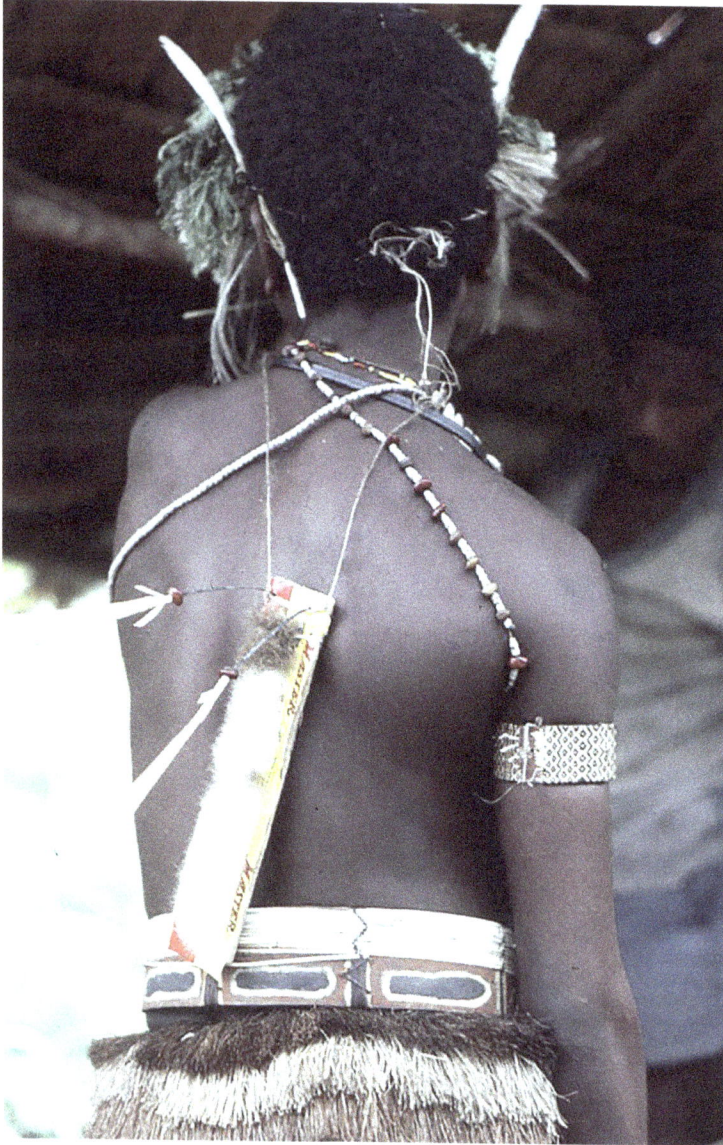

Figure 2.3: Young man, decorated in preparation for a feast.
Source: Photograph by Peter D. Dwyer, 1986.
Note: A label from a can of fish has been used in the decoration.

By 1986, people at Gwaimasi were already aware of the new foods that white men carried and ate. At the 1986 feast, one man from Nanega added a kilogram of rice and a tin of fish to his prestation; we provided the same to Monu, so that the gift could be appropriately balanced. Over the next few years, rice and tinned fish became standard additions to feast fare. During a serious drought in 1997, the Australian Army delivered supplies of rice, tinned fish and oil to the community as part of a nation-wide relief effort; relief that was not, in fact, needed in the Gwaimasi area (Minnegal and Dwyer 2000a). But such 'modern' foods had not become a regular part of the diet for anyone at Gwaimasi by 1999.

The World of Spirits

In small communities where the acquisition of resources cannot be guaranteed, sharing is necessary. It buffers returns. Sharing food was an everyday occurrence at Gwaimasi. A woman returning from processing sago would give a portion to someone in a neighbouring family and might receive in return bananas or, indeed, a portion of sago equivalent to the one she had given. A man returning from a successful pig hunt would deposit the carcass at someone else's hearth and retire to his own house. Others would emerge, butcher the animal, light a fire, heat stones and make an oven. The pig belonged to everyone and when it was cooked, men, women, children who were not nursing and anthropologists would each bring their own plate and receive a share. At Gwaimasi, people shared in the food resources to which, as residents of that community, they held collective rights. And they shared out resources that they themselves had acquired to some or—if the amount was large—all members of the community (Minnegal 1997).

But sharing food does more than merely buffer returns. It creates goodwill. At Gwaimasi, people were deeply concerned with sorcery. At night they were vulnerable to *we*, sent by mountain people, which travelled along water courses to seek out sleeping people. These spirits were often held responsible for the deaths of infants or for epidemics that spread through communities. When many people at Gwaimasi fell ill in 1987, suffering from what we interpreted as influenza, and two older men appeared close to death, Biabia—a *soi*, able to see the activities of such spirits, though not directly communicate with them—was summoned from the next village to diagnose the problem. He removed the tiny bamboo 'arrows' that had been shot into these two men, and advised that they retreat to bush houses to hide from further attack. In the forest, people were vulnerable to assault sorcerers—*hugai*—which were thought to come from the lands of Pare people to the southwest. *Hugai* killed people, stole organs, returned the victims to life and allowed them to return to their own houses where they soon fell ill and died. And people were also vulnerable to parcel sorcery—*bogei* sorcery—which someone, either from their own or a neighbouring community, could initiate by stealing hair, nail clippings or possessions identified with a particular person and, using these, call on dangerous spirits to harm the chosen victim. People at Gwaimasi shared food with others to show that they themselves did not have sorcery in mind, and to offset the risk that they themselves might be ensorcelled.

There were, as well, other spirit beings that could aid or harm people. The spirits of the dead resided for a time in the bodies of animals but ultimately rested at a *toi sa*—a forbidden place—that could be visited only by bachelors, and then only after at least one of them had been purified by dancing through the night. The spiritual essence of living people left their bodies at night and these spirits too could cause harm. When Geswa fell ill, the visiting spirit medium Tobasi conducted a séance in which he communicated with spirits, learned that Geswa was being attacked by a crocodile spirit and, with his apprentice son, oversaw a curing dance at which the implicated spirit was challenged and evicted from Geswa's body (Dwyer and Minnegal 1988). When Tobu was diving for fish in foothills east of the Strickland River he saw a large catfish—*Aiodia* (*Plotosus papuensis*)—outside its usual altitudinal range. It harboured a spirit being that was guiding its movements. Tobu's health, and that of his younger brother Sidine, was threatened. A curing dance was deemed necessary. Yakabai danced through the night, painted and costumed and wearing a cassowary bone dagger through his arm band. When the butcher bird called at dawn the dance finished. A screaming piglet was held taut above Tobu and Sidine. The now spiritually-strengthened dagger was plunged into its body and its blood ran onto the two men. Parts of the piglet, together with a powerful spherical stone—an *ugwi*—were buried in the forest and the danger to the two brothers passed. There are other beings too. A python without a tail guards the *toi sa*, the huge eel Kongwa is found in the river and larger streams and a giant hunter, Nakobaia, with multi-segmented limbs and a corkscrew neck, roams the back swamps with his bow and with arrows that never miss. For Kubo, the invisible world of spirits permeates the land; it is coextensive with the visible world (Dwyer 1996).

By 1986, two further spirits—Godi and Yesu—had been added to this pantheon, and prayers to these were often said before meals, before journeys, or when someone was ill. People had learned of these beings from the pastor at Sesanabi, a few years before, but they had been added to, rather than displacing, the spirits that populated their world. People at Gwaimasi had no doubt that Godi and Yesu were powerful, but they were not of the forest; and, it seemed, while they could be appealed to, people were unsure how to know what they wanted or what they were doing. One young man from Gwaimasi, Gehogwa, had been sent to school (five days walk away) so that he could learn to read the bible and reveal its message. In 1986, after four years of schooling, he occasionally held prayer meetings and bible readings during breaks from school. But there was no pastor within a day's walk of Gwaimasi, and it was not till 1995 that a church was built in the community and regular weekly services held. By this time, séances and curing dances had been abandoned, and it was now through prayer that people sought some control over their fates.

Expanding Horizons

In 1986–87, Gwaimasi was a small and materially, though not socially, self-sufficient community. Admittedly, steel tools such as axes and bushknives had largely, though not entirely, replaced stone tools (Fig. 2.4).[8] Most people had metal cooking pots, plates and spoons, and people—men more than women—had at least some Western clothing. Salt and soap were now accessed through distant trade stores and fishing line and hooks had not been available until Europeans arrived in the area. But at this time it was, perhaps, only the axes and bushknives that had become necessities. For the most part, what was needed to build houses, make clothing or the tools wanted for gardening, sago processing and hunting could be, and most often was, acquired from the nearby forest. To get these necessities, however, people moved a lot.

---

8   For some tasks, people at Gwaimasi preferred stone over alternative materials. In particular, the pounders used in processing sago were made from knapped chert cores. By contrast, steel greatly reduced labour costs in the construction of canoes though the most appropriate tool was not easy to access and, by local standards, expensive. In 1986–87, the residents of four villages—speakers of four languages—spread out along 40 km of the Strickland River shared use of a single canoe adze.

Figure 2.4: Young man sharpening an arrow with knapped stone.
Source: Photograph by Peter D. Dwyer, 1987.

In 1986–87, Gwaimasi residents spent nearly a quarter of available nights at a garden house, or at a house from which they hunted or processed sago, scattered through a 50 km² area that we refer to as the local subsistence zone (averaged over a 14-month sample period; Minnegal and Dwyer 2000b). Females were slightly more likely to be absent than males (24 versus 20 per cent of available nights, excluding children). These absences might be for one night only but, more often, were for several days or even a week or more. A man with his wife, and children if there were any, might leave the village to fell a sago palm and process the flour. It was usually women who washed out the pith though there were men who sometimes helped with this work. More often, however, the husband fished or hunted while his wife processed sago. At other times a woman accompanied her husband when he went hunting, perhaps sitting at the base of a tree to catch birds that fell after he had shot them from a hide concealed in the upper branches. And sometimes, if a feast was pending, a group of men with dogs would travel together and hunt for pigs and other game in the daylight hours.

The distances people travelled were not great, often just a few kilometres from the village. With garden and sago work they could have returned each evening. People stayed away because they enjoyed their time in the forest and, perhaps as well, because it obviated the not particularly arduous need to share. What they ate while away was of huge interest to everyone. When seven-year-old Agiyai and three-year-old Kaibo returned from some days away with their parents they were always called to someone's hearth, given food, and asked 'what did you eat?' It was the meat foods that people wanted to know about. We, of course, did the same.

There were, as well, many movements beyond the subsistence zone. People visited a Kubo community at Sosoibi, on the east bank of the Strickland River, a journey of one hour to the south by dugout canoe or raft, and later, when these people relocated, visited them at Gugwuasu six hours walk to the southeast. They visited a Konai community at Nanega, three hours walk to the west, and a community of Konai and Febi people at Ogwatibi, a day's walk to the north. These visits were for social reasons. People visited kin, attended feasts or funerals, joined hunting ventures, negotiated marriages, resolved disputes, or participated in séances and curing dances at which illness was diagnosed or treated. Males spent 9 per cent of available nights at these communities or within the subsistence zones associated with them; for females the equivalent value was only 3 per cent. Hunting and the politics of dispute resolution and séances account for most of this gender disparity. Other movements—11.5 per cent of available nights—were further afield, entailing more than one day to reach the destination. In these cases, however, the primary rationale for travel was 'modern' in that most trips were to the community health centres at the mission stations of either Suabi or Dahamo and were undertaken because infants were ill.[9] On another occasion, some men and women travelled to Dahamo to vote in the national election, though they knew little of either the candidates or the issues. When one youth, Gehogwa, left Gwaimasi to return to school near Kiunga, five days walk away, he was accompanied by his older brother-in-law who remained away for a month. There were no other movements of this magnitude.

In the longer term, based on records across ten years, people often relocated from one place to another because the site of a particular hamlet changed or, more often, because marriages and disputes were associated with the departure of either wife or husband to join the community

---

9   At Suabi Mission Station an airstrip was opened in 1984 and a community health centre soon afterwards. At Dahamo Mission Station these facilities arrived in 1987. Gwaimasi was closer to Dahamo than to Suabi, and when the latter community health centre opened people switched their allegiance to it.

of their spouse or with realignments of entire families (Minnegal and Dwyer 2000b). Departure was the conventional way in which Kubo people and their neighbours resolved disputes: to a bush house for a short period if the dispute was minor, by relocating to another community if the difficulty was greater. The knowledge that you were suspected of sorcery could provide the motive for relocation. What was striking, however, was that over a ten-year period people who had left a community often later rejoined the original group. The attraction of their *oobi* land was very strong. In the short term, people appeared to be very mobile, but in the longer term small communities or, better, small sets of communities were remarkably cohesive. That coherence arose from a network of agnatic and affinal connections, together with those of initiation, that tied people together. It was that coherence that prompted us, in an earlier report, to depict Kubo as 'sedentary nomads' (Minnegal and Dwyer 2000b).

With a growing recognition of facilities available only elsewhere, people at Gwaimasi were expanding their field of movement. They carried infants to health centres, sent children to schools, travelled to distant trade stores to purchase clothes and tools. While movements had previously been shaped by social imperatives, these 'modern' movements followed individual agendas. The health of a particular person or the acquisition of new knowledge by a particular person were no longer to be assured by appropriate relationships within the community, and did not in themselves enhance relationships. The survival of an infant, however, kept open relational possibilities. Local relationships continued to be focal for Kubo. Gehogwa returned to Gwaimasi in late 1991, after completing Grade 10 studies at Rumginae and Kiunga. The children who attended schools that opened at Suabi in 1987 and Dahamo in 1988 also returned. Clothes and tools purchased at distant trade stores were brought back to the village and shared with others. Horizons had expanded, but reaching out towards these was done, it seemed, in an effort to draw the resources of that wider world into the ambit of local communities. Simultaneously, however, people were being drawn out of—disembedded from—their place; children at distant schools no longer learned to read the land, and Gehogwa was not allowed to forget that he had been away during the initiation at which his peers at Gwaimasi had become men.

Leadership and Dispute Resolution

There were no acknowledged leaders at Gwaimasi, no one to whom others accorded the authority to speak for them, or to instruct on what should be done. Admittedly, one man did have a title, specifying a role and status that existed beyond the incumbent. In 1986, Monu was often addressed as 'Komit'—not a name, we eventually learned, but a title that denoted his position as the government-appointed 'Committee' for the local community. But there were no tasks, and no power, associated with the role. Monu did not attend meetings at distant centres. Nor did he convene meetings or take a leading role in those that did occur in the community. Some people were acknowledged as skilled in particular domains: Uhabo made superb arrows, Daledi killed far more wild pigs than other men, and Fafobia composed songs that everyone admired. But this did not give these people special standing in domains beyond those in which they had distinguished themselves. Indeed, since performing a task particularly well one day did not necessarily guarantee effective performance on future occasions, their authority even in those domains was not generally acknowledged.

To say that Kubo did not recognise authority does not mean that no one had any influence over others—that there were no leaders. But influence was exercised through action not instruction. In 1995, we watched as Gehogwa, now graduated from high school, sought to establish a small crocodile farm to bring in some regular money to the community. He did his homework, contacted buyers and obtained advice on how to build pens. Then he called a meeting and told everyone what needed to be done. But no one did anything. The project never got off the ground. In contrast, in that same year, Daiwo—a 50-year-old man once resident at Gwaimasi but now visiting from his own land to the north—decided that the village should be cleaned before a pending feast. He said nothing to anyone, but one morning at dawn began slashing the grass. Others emerged from their houses, watched for a while, then gradually joined in. And within an hour Daiwo was sitting on our veranda, watching, as those others continued the work he thought should be done. Each of those others had independently decided that it would suit them, too, to have the village looking good when the visitors arrived.

But, as use of the title Komit implied, people at Gwaimasi were not unaware that new sources not just of moral leadership but also of power were becoming available—sources that, like Godi and Yesu, lurked at the margins of their world but, at times and with the right knowledge, could be drawn on to shape events.

In January 1987, the young unmarried man Hami asked us to write a letter on his behalf. Gufu, in her late teens and unmarried, was pregnant and it was thought that Hami was responsible. To the people at Gwaimasi, there was 'no road'—no relational future—for a child born out of wedlock. Gufu's stepfather and her half-brothers threatened to kill both her and Hami and to throw their bodies in the river. The letter that Hami dictated was addressed to the police at the government station at Nomad, three days walk away. Hami explained the problem and asked the police to jail him. He did not ask that they protect Gufu. Though he asserted his innocence he felt that incarceration would both remove him from the immediate threat of death and, eventually, be accepted as a satisfactory alternative punishment for the imagined wrong.

For two days there was much animated and often angry discussion. On the evening of the second day Hami, Gufu's stepfather and a few other men came to our house. A decision had been reached and, with few Tok Pisin speakers at Gwaimasi, it was now Hami's role to translate. They had decided not to kill him, being concerned that if they did so we would be frightened and might return to Australia. Hami kept, but never delivered, his letter.

Twelve years later, in January 1999, we observed a lengthy but failed attempt to resolve a complex dispute that involved Kubo, Konai, and putative Awin-speakers (Fig. 2.5).[10] A few weeks earlier, the young man Dengwa had married Biwo, the daughter of Biabia's sister who had married an Awin man but whose marriage had not been directly reciprocated. As a young girl, Biwo had left her home village and commenced living with Biabia's family. It was intended that she would eventually be an exchange sister for one of Biabia's sons; in effect she would replace her mother in Biabia's lineage, thus negating the imbalance the earlier unreciprocated marriage had caused.

---

10   The Awin-speaking participants in this dispute were referred to by Kubo people as Habiei and were the survivors and descendants of a former Kubo *oobi* who had, in the late 1960s, relocated to Awin territory after an attack by other Kubo which resulted in several Habiei men being killed and eaten.

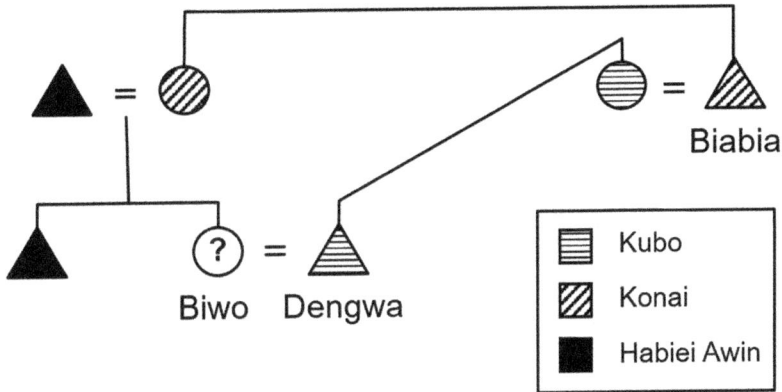

Figure 2.5: Key participants in a dispute about marriage.

Source: Authors.

Note: Shading shows *oobi* affiliation.

In fact, Biabia had no biological sons. Dengwa was Biabia's wife's younger brother and, in part, Biabia constructed the Dengwa–Biwo marriage as balancing his own marriage to Dengwa's sister. That is, he fabricated Biwo as his own sister rather than as a sister to his son.[11] No one, however, informed Biwo's Awin relatives of what was happening. They learned of the marriage only when Biwo's biological brother arrived from his Awin village to claim his sister. He too wished to marry but his intended spouse's brother would not sanction the marriage unless it was immediately reciprocated. He came with a substantial contingent of other young Awin men. The Awin did not accept Biabia's assertion that Biwo qualified as the substitute exchange sister for his own sister. Rather they argued that the marriage of Biabia's sister to an Awin man had been in compensation for one of five cannibalistic deaths that had occurred in the late 1960s and, thus, no reciprocation or substitution was required. There were hours of discussion that were forceful but polite. But neither side moved from their firmly held position. There was no resolution. In the end Biwo's brother said that he would take the matter to 'court' at Nomad. Biabia said: 'Go ahead. I will be there'.

---

11   The Dengwa-Biwo marriage was also constructed as an immediate sibling exchange. Dengwa married Biwo at the same time that Awabu, who on the basis that his early years were spent in Awin territory was constructed as a classificatory brother to Biwo, married Dasogo who was a relatively unambiguous classificatory sister to Dengwa. If Dengwa's marriage was broken then the validity of Awabu's marriage would have been also in doubt.

In the years to 1999, relatively few disputes surfaced among the people at Gwaimasi and neighbouring communities. If a married couple quarrelled, two families fell out or there was concern about sorcery then, for a time, one of the protagonists would leave the village. When Tobu shot his sister's dog—to hurt but not to kill—because that dog had killed his caged young cassowary, an open discussion, facilitated by a senior man, ensued and Tobu and Fafobia were each required to give PGK20 to the other. The amounts were identical but each of them delegated someone to carry their 20 kina note to the person they had offended. The notes changed hands in public and simultaneously. The two disputes summarised above were, however, judged to be more serious. And, in both cases, one or more participants sought to deflect potentially serious and immediate consequences, or defer outcomes that they did not desire, by appealing to the higher authority of governmental structures and processes. In neither case, however, was that appeal directly implemented by visiting Nomad. Had this happened it is unlikely, as Bruce Knauft (2002b) made clear in *Exchanging the Past*, that administrative officers based there would have been particularly interested in adjudicating or have had the means to adjudicate in relation to such complex disputes.

Northern Kubo people living near the Strickland River had, and continued to have, remarkably little direct engagement with the paraphernalia of law and order associated with either colonial or national governments. If problems arose they were expected to walk to Nomad and air their grievance; the police did not come to them. What happened instead, with some disputes, was that people strategically employed imagined structures of external governance to stall or deflect possible outcomes of customary modes of dispute resolution where they judged the likely outcome to be unfavourable to them. In doing so, however, they disrupted conventional forms of resolution—departure, reciprocal exchanges and, in serious cases, assault that might lead to death—that were embedded in a relational ethos. In the case of Hami, for example, his decision to write to the police, and to seek incarceration for an offence that he may not have committed, was made without reference to the opinions of other members of the community with which he was affiliated. He did not even seek protection for the young woman Gofu. In a context of opportunities that were imagined to be provided by the bureaucratic structures located at Nomad, Hami had individualised decision-making and, to this extent, negated the pre-existing and predominant relational basis of dispute resolution.[12] Again, as it became clear that there was no easy resolution to the complex dispute arising from Dengwa's marriage, and thoughts turned to Nomad, the terms of reference were collapsed and consolidated as involving only Biabia and Biwo's brother. Here again, therefore, issues of blame and morality were being disembedded from a network of relational others as the epistemological foundations of Kubo lifeways shifted subtly to a prioritisation of the categorical and existential. An erosion of the relational ethos and a heightened emphasis on the individual characterised many changes that occurred among the people at Gwaimasi through the years that we visited them.

Money and the Commensurability of Difference

When we arrived at Gwaimasi in August 1986—via three helicopter trips from the mission station at Suabi—we brought supplies: personal items such as clothing, bedding and toiletries; a medical kit for ourselves and to tend, where we could, local needs; work items such as tape measures, balances, stationery; a solar panel, battery and lights; some tools and food and money. We needed money to pay for our house and firewood, for other occasions when we employed people, to buy food and, because we planned to monitor hunting, to buy the skulls of fish and other animals. For food and skulls we needed coins. But coins, in bulk, are heavy and in PNG air freight is expensive. So we carried less than we needed intending that, as our supply diminished, we would exchange paper money for the coins that people had accumulated. At the outset there were few people who had any money at all and for some months our contributions were greatly valued.[13] The time came when our own supply of coins was low

---

12   In 1986–87, Hami's instances of individualising decision making were more frequent than those of most other Kubo we knew well. In his early teen years he had accompanied an SDA pastor on patrols through Samo and Bedamuni territory and spent some time at an SDA mission and teaching centre near Komo, in Huli territory. Thus, during his formative years, he had greater exposure than most of the Kubo people we knew well to different understandings and different ways of living.

13   In the 15 months from August 1986 the total income to Gwaimasi was between PGK1,700 and PGK2,000. Our presence, combined with a brief visit by representatives of a company that had been exploring for gold in the region and came to pay for ongoing maintenance of the camp they had previously established about a half-hour walk north of the village, increased people's access to money. Through a 399-day period, each person older than about 15 years earned an average of between PGK85 and PGK100. Males spent on average 22 days (5.5 per cent of their time) and females spent on average 4.3 days (1.1 per cent of their time) in tasks specifically directed at earning money (Minnegal 1994: 102–6). Through the early months of the survey nearly all money received was spent—most

and we asked people to bring us theirs to exchange for notes. No one responded. More time elapsed and our supply seemed likely to run out. We became more insistent. People felt sorry for us, and sought to oblige. Someone might carry PGK2.30 in coins, and in return we would give a PGK2.00 note and 30 toea in change. Someone else might carry 80 toea, and we would give it back; there was no exchange. It seemed to those who were trying to oblige us that we were thoroughly inconsistent. Sometimes we took some of their money and returned the rest. Sometimes we rejected their offer of assistance. We needed to find a solution.

We had learned early on that when people talked about money they held or had spent they spoke of coins or notes rather than an aggregated total. A person who had visited the mission station at Suabi and bought a bushknife would explain that it had cost, for example, 'two kina three, one kina one, twenty toea six and ten toea three'. They would not say that the bushknife cost PGK8.50 and, indeed, were likely to speak of '*beio muni*' (cuscus money) or '*djiwo muni*' (cassowary money) in reference to the animals depicted on, respectively, 10 toea and 20 toea coins rather than invoke either the abstract value of a coin (10 or 20) or its abstract standardisation (toea or kina) as a unit of currency. It was not that people at Gwaimasi could not count or, to some extent, add. It was simply that they did not know that two *beio* was equivalent to one *djiwo*. Without realising what we were doing, we proceeded to teach them what they did not know, and thus contributed to changes that swept through their society for the next decade and more (Minnegal and Dwyer 2007).

Our solution to our problem was to prepare a chart on which we made pencil rubbings of all combinations of coins that yielded one kina and to reinforce the understanding that two one kina coins had the same value as a two kina note. Now people would bring their coins, lay them out on top of our pencil rubbings and provide us with sets equalling one kina. It was through this process that people came to appreciate that different denominations of money were commensurate. Where, before, particular coins or notes were personalised—the history of each was known and could be recalled—people came to appreciate money as reified, as an

spent on clothes, some used to purchase axes, bush knives, fishing tackle and cooking pots, and some to buy small quantities of rice and tinned fish that were served at intercommunity feasts. After about eight months, however, most people had received as much money as they currently had a use for. From that time on they were less willing to undertake more time-consuming tasks despite the fact that our pay rates were considered to be relatively high compared to those of missionaries or exploration companies (Minnegal and Dwyer 2007: 209–10).

abstract category. It was this sense of categorisation that increasingly, though never absolutely, came to permeate people's understandings of and relationships with pigs, women, assemblages of people and land.[14]

Pigs

In 1986–87, when Kubo people exchanged domestic pigs the ideal— and most often the practice—was that the exchange occurred between two people, not between groups, and that the pigs were of the same size, the same sex and the same colour.[15] Like was exchanged for like. Further, it was often the case that the two pigs had been captured by men when they were wild piglets. These men would each give their piglet to a woman—often a classificatory sister but not necessarily their exchange sister—to care for through the next few years on the expectation that, in the future, they would receive that pig from the woman's husband. In the years immediately following 1987 people learned that schoolteachers, community health workers and, occasionally, visiting government or company employees would sometimes buy domestic pigs. Through those years too their desire for money, and indeed their conceptualisation that money was needed, increased. Pigs, they decided, were a means to satisfy their desire and need. Thus, though in fact there was no market, they committed a great deal of time and effort to increasing the size of their holdings (Minnegal and Dwyer 1997). By 1995 the domestic pig population at Gwaimasi had more than doubled. In the earlier period, in any one month, there was an average of 0.38 domestic pigs per person; in the later period the average was 0.88. Now, people would have sufficient pigs to take advantage of any offer from outsiders to buy one, while still meeting local social obligations. Now they were prepared to

---

14   Gwaimasi and its neighbouring communities close to the Strickland River were remote. Government officials had seldom visited them in the years before Independence (1975) and, to the best of our knowledge, never in the years after Independence. While, at Gwaimasi, we contributed to an appreciation of the commensurability of difference, people living nearer to the patrol post at Nomad or mission stations in the region will have been introduced to the same notion in other ways and, probably, somewhat earlier. Nor is it unexpected that it was more than 20 years after Europeans first came to the region that Kubo people began to incorporate and act on this notion. Salisbury (1970: 180–7), for example, observed that it was at least 25 years before Tolai people of East New Britain accepted cash payments and that, 85 years after European contact, he himself reciprocated 'gifts' of food with tobacco rather than with money because the former was understood as 'barter' between friends and the latter would have been an insult (Minnegal and Dwyer 2007: 17).
15   As reported by Jadran Mimica (1988: 20, n. 8), Thomas Ernst noted that among Onabasulu of Mt Bosavi, Southern Highlands, 'not only is there a concern with the exact size of two exchange piglets, but it is made sure that they have the same sex and, in so far as it is possible, exactly the same fur colour and spots'.

forego the expectation that all exchanges be of like for like and, rather, to accept a certain amount of money as payment for a pig of a certain size. In these contexts, a pig was just a pig. The particularities of its origin and ownership, and of the care given by one woman over a long period—particularities that made visible the relationships between particular people—were no longer relevant to the hoped-for transaction. They were of no interest or consequence to the person who might buy the pig. The pig was suitable for purchase simply because it was a 'pig'. For people at Gwaimasi, the boundary between animals categorised as 'pig' and those in other categories was now more salient in guiding social action than were the relational particularities of pigs that had hitherto affirmed them to be, or not to be, appropriate to a specific exchange.

An external market for pigs failed to materialise. Indeed, through the later years of the 1990s the failure of government to fund remote schools and health centres saw the withdrawal of government employees and, hence, even less likelihood that a schoolteacher or community health worker might buy a pig. Yet by 1999, people at Gwaimasi kept even more domestic pigs with the ratio of domestic pigs to people now averaging 1.0 (Minnegal and Dwyer 2000a: 507). By this time, however, in the absence of the expected market, Kubo people and their Konai and Samo neighbours began acquiring pigs for events such as church feasts and marriages by buying from each other, though only from those of a different language group (Minnegal and Dwyer 2007: 15). The recognition and acceptance of 'pig' as a category had been extended to people who spoke different languages. Exchanges between Kubo people themselves remained embedded in the relationships that tied those people together. It would be years before these too were disrupted.

Women

For Kubo people, immediate exchange was the ideal. As seen above, it applied when pigs were exchanged. It also applied in the case of marriage, where two men would exchange sisters as brides—though always on the proviso that the women themselves agreed to the transaction.[16]

---

16 As the paired marriages of Sinage and Sidine, and of Uhabo and Tobu, illustrate, it was not always possible to achieve the ideal of immediate sibling exchange. A common and continuing strategy in cases where a man has no sister, or a couple have more sons than daughters, is for parents of the wife to adopt one of her daughters as a future exchange sister to one of their sons (Minnegal and Dwyer 2006: 131–2, n. 6).

By and large, in 1986–87, marriage negotiations concerned only these four parties. Indeed, because few people lived beyond 50 years it was often the case that the parents of the prospective couple were no longer alive. The primary constraint on eligibility with respect to marriage was that the man and woman were not to be of the same *oobi* or, in fact, of *oobi* that were said to be 'brothers' because their lands were either contiguous or linked mythologically by, for example, underground passages. There were cases where this constraint was ignored (the couple eloped) and these were routinely the subject of disapproving gossip. And there were other cases where a couple pre-empted negotiations with exchange siblings, eloping before an exchange had been agreed or even when an exchange had been rejected; these cases, too, elicited much critical gossip, and intense debate about how the de facto marriage might be reciprocated and thus legitimised. But always, as with customary exchanges of pigs, it was the relational particularities between individuals that were foregrounded and validated the marriage. It was not that a 'man' married a 'woman'. It was that a particular man, embedded in a particular set of relationships, married a particular woman who, likewise, was embedded in a particular set of relationships. It was their union, put in place by their actions, which established or reinforced affinal connections between the *oobi* to which they belonged. It was not pre-existing connections between those *oobi* that served as the backdrop to the marriages. Agency, rather than structure, was primary.

In 1986–87 the relationship between a husband and wife was extremely close. The family was an autonomous unit, gardening together and, very often, spending time together in the forest or backswamps when the husband was hunting or the wife was processing sago. Public displays of affection were not uncommon. A newly married couple might walk through the village holding hands, or a man might lie with his head in his wife's lap while she searched for lice in his hair. But this began to change through the next few years, as people increasingly looked to those bearing word from the wider world for guidance as to how the opportunities evident in that world might be accessed. New forms of Christianity arrived with messages that clearly differentiated, and differentially valued, men and women. Initially, all Gwaimasi residents were affiliated with the ECPNG and, though there was no resident pastor, Sunday services were not uncommon.[17] In January 1987, Tiotidua came from Suabi in marriage to Geswa. She was affiliated with the SDA Mission and, thus, was not to eat fish without scales, turtle, bandicoot or pig. At first she complied with these dietary strictures though, in compensation, people were very generous in ensuring that, when they ate what she couldn't, she was given meat that she could eat—particularly fish with scales and birds. Gradually, however, Tiotidua drifted away from these constraints by, at first, eating fern leaves that had been cooked with pig and were enriched in fat and later, after about six months, sharing pig with everyone else and aligning with ECPNG. The different church paths and practices, and the different ways these divided up the things of the world as edible or not, were less salient to people at Gwaimasi than the potential benefits offered by affiliation with a Christian mission. Through all this time, people considered that these missions had contributed more to their well-being than government. But they wanted more; in particular, they wanted an airstrip and their own white missionary.

By 1995, some disillusion was apparent. A Konai pastor had lived at Gwaimasi for a year or so but, with his family, had departed. There was no airstrip. White missionaries had not come. People told us that ECPNG was 'not strong' and, for a time, were refocusing attention on the Christian Brethren Church which was actively proselytising among Febi communities to the north and extending its reach to the south. Huli pastors had visited and one charismatic Febi man, who at this time was often present, held prayer meetings, led hymns, asserted that assault sorcerers were common near the village, that the world would end in the year 2000 and that Jesus's return was imminent (Minnegal and Dwyer 1997). Drawing on *Genesis* he taught that it was women who were responsible for the downfall of men, held private meetings with women where he instructed them in proper behaviour and, with considerable success, insisted that women should, at all times, cover their heads. Church services became more frequent as this man progressively imposed a structural separation between men and women—a separation that was increasingly played out in more secular domains of life too, because the increased number of domestic pigs had simultaneously increased the work of women, tied them to the village to a greater extent than before and greatly reduced the likelihood that families went to the forest together.

---

17   Services were held in most weeks from the time of our arrival in August 1986 to late January 1987. During this period the youth Gehogwa officiated but, when he returned to school, the frequency of services was greatly reduced.

Husbands and wives were now far less likely to be together than in 1986–87. The impact of this phase and form of Christian expression was to subsume particularised social identities as sister, wife or mother and as brother, husband or father within overarching and contrasting categories of female and male. Again, therefore, in the years after our first long stay at Gwaimasi, a relational epistemology was giving way to a categorical epistemology.

People and Land

At Gwaimasi, an emergence of categories as salient where they had not been so before was manifest in other ways too. By 1995, when small feasts were held, food was no longer distributed to each person but rather to 'groups' which, at that time, comprised those people of, and most closely aligned with, each of the two primary *oobi* that constituted the community (Dwyer and Minnegal 1998: 34–5). Married women associated with their natal *oobi* rather than with that of their husband. At larger feasts too, where people from other communities attended, groups rather than persons were now focal to the distribution of food, with the two Gwaimasi groupings treated as distinct.[18]

In 1986–87 neither rights to land nor use of land were congruent with *oobi* membership. Many people were resident on land other than that which was identified with their *oobi*, and here, though at least with sago they contributed less work per person than did those of the *oobi* whose land it was (Dwyer and Minnegal 1997), they were relatively free to make gardens and to take forest products as required. Indeed, though people did not hesitate to name the *oobi* to which they considered they belonged, this had little salience in everyday life. *Oobi* membership did not, in itself, dictate conventions of resource acquisition and marriage. Rather, it arose out of practice, out of the recognition that they were members of a group who obtained the wherewithal of subsistence from shared land and who did not intermarry. Moreover, people did not identify particular bounded areas of land as being either their own land or the land of someone from another *oobi*. In the first place, they spoke of land as associated with a particular person, rather than a group of people. And in identifying someone's land they spoke of focal sites—the junction of two streams,

---

18   The shift to 'group' level sharing proved a little embarrassing for us in that, early on, we were identified as a separate group which, despite having a membership of only two, should receive as much as other groups.

a cave, a named hill, the site of a former longhouse. They did not refer to boundaries. It seemed that the associations to which they referred, and their sense of belonging, decreased in intensity as they moved away from any one of these sites. And, in the second place, they rarely spoke of these sites, or of the areas around these sites, as being identified with their fathers or with themselves. Rather, if they were married and had children, they named those sites or areas as belonging to those children and, in this way, wrote their children onto the land. With respect to land, and in other ways too, genealogical connections were submerged—indeed, few people could trace back more than two generations—in favour of an orientation to the future, to those who would come after them (Minnegal and Dwyer 1999: 65–6). In those years, then, social identity arose in an always fluid context of relationships with a particular set of people who lived on, and used, a particular area of land. And it was given expression in action that was shared with those people and that land. It was only when referring to peripheral people or places—to their deceased forebears, to people who lived at a distance on unfamiliar land—that '*oobi*' or 'language' identification became salient. Categorisation, in these ways, was foregrounded only in the absence of ongoing interaction.

By 1995 there were signs of change. Rumours had reached Gwaimasi that government was planning a country-wide register of land ownership.[19] There was concern that land which was not used, or could not be shown to be 'owned', would be acquired by government. The matter was the subject of much discussion and, for the first time in our experience, one focus of that discussion concerned 'borders'—marks or lines on the ground that separated the land of one *oobi* from that of another. The English words 'border' and 'mark' were deployed in these discussions; Kubo, it seemed did not have suitably analogous terms.

For the first time, too, a dispute about land—or, rather, about the use of land and about appropriate places of residence—emerged (Minnegal and Dwyer 1999). The village of Gwaimasi had been in place for more than nine years. In customary practice longhouse communities had moved every three to four years. Now, however, as a consequence of longer-term residence, convenient locations for gardens were more distant from the village. People continued to eat well, for there was no diminution in available bush foods, but it proved more difficult to feed visitors

---

19   At that time, an offer of a loan by the World Bank to PNG was contingent upon development of a register of customary land-ownership. The register did not eventuate.

appropriately at feast times. On these occasions, it should be evident to everyone that the food provided embodies the productive effort of the hosts, and bush foods were not the product of human labour (Minnegal and Dwyer 2001: 282). Further, residents who were not of the *oobi* on which the village had been established were now using resources—especially sago—more freely than had previously been the case. They were acting as though long-term residence gave them the right to do so without a need to either wait or ask to be invited.

It was in this period that a young man, Sigio, wanted to marry Agiyai. He was of the *oobi* on whose land the village was established. He had no sister. While by this time there was some acceptance that sibling-exchange could be replaced by bridewealth, Sigio also had neither money nor kin who could provide monetary support. His offer of marriage was rejected. He was angry, and demonstrated this by starting to chop down a coconut palm that belonged to Agiyai's mother's brother. He shouted to those who gathered to watch, asking what the outsiders—the people of other *oobi*—were doing here, using resources that were not their own. He asserted that he would 'spoil' the village and force those others to leave, to return to their own lands.

Sigio's desire to marry Agiyai did not come to fruition. Nor did he succeed in felling the coconut palm. In expressing his anger, however, Sigio appealed to *oobi* identity, to the possibility that *oobi* were corporate groups and held inalienable rights of ownership, rights that were given prior to either residence or the ongoing and productive use of land which would be visible to everyone. And in the action that he initiated—his attempt to fell a coconut palm—he sought to remove from 'his' land a material expression of the palm's owner and, thereby, cut the network of relationships that that man and his kin had, through their actions, established both with the land and with him. Sigio had, as we observed with both pigs and with gender, de-emphasised relational imperatives in favour of categorical ones.

Changes in the layout of the village itself reflected the same shift in emphasis. In 1986–87, the longhouse at Gwaimasi was available to all, the place where visitors to the community were hosted, fed bananas and tobacco while exchanging news from elsewhere. The small family houses people built nearby were open, dominated by a large unwalled cooking area on ground level at the front. A smaller walled sleeping area was accessed by a few steps up from the sitting platform at the rear of the hearth, but the door to this was closed only when those living there retreated inside for the night. People moved freely back and forth between these houses, carrying gifts of food, sharing what had just been cooked, and gossiping. Indeed, the houses themselves were close together and neighbours could converse without moving from their own hearths. By 1995, however, there was no longhouse at Gwaimasi and family houses were larger, more widely spaced and more solidly built than before (Dwyer and Minnegal 1998: 34–5). Houses were now fully walled, and hearths were inside behind relatively secure doors. People no longer wandered freely in and out, even when the door was open, but usually knocked and waited to be invited inside. And visitors to the community now had to decide whose house they would go to on arrival. Sometimes they chose ours, for we had retained the earlier structural arrangement.

The size and solidity of these new houses, the closed doors and the isolation they imposed, even the internal partitions that assigned individuals within the household their own private space, reflected an increasing emphasis on differentiating people by their possessions rather than by the relationships they had established with others. Pigs were increasingly valued according to size or colour rather than who had reared them, women according to age, attractiveness or fertility rather than whose sister they were, and men according to *oobi* identity rather than what they had planted or built and the places where they had done these things. So, too, persons were increasing valued in terms of the things they accumulated—things others might desire—rather than the connections those things mediated.

By January 1999, with the exception of one family, Gwaimasi was abandoned. People had returned to Mome Hafi where, about 35 years earlier, Sinage, Monu and Tobu had been initiated. Now, however, they were being joined by a river-side community from the south. New houses were under construction. A church had been built but people had not continued their earlier flirtation with the Christian Brethren Church. The new village would include 53 people together with a Samo-speaking ECPNG pastor, his wife and four of their children. Sinage, Uhabo and Tobu continued their close residential alignment but Monu, who had left his first wife and remarried, had moved to Suabi. Sigio's expressed desire that people should live on their own *oobi* lands had not come to pass though he himself did not make the move to Mome Hafi but, restlessly, lived sometimes with his half-brother's family at Gwaimasi and, at other times, alternated between neighbouring Konai and Febi communities.

He was not yet married. At Mome Hafi, however, there continued to be concerns about the use of land, concerns that were now expressed by a claim to ownership (Dwyer and Minnegal 2007).

One of the residents of the new Mome Hafi community was Noah who, with his wife and daughter, had relocated from Suabi. Noah was the sole surviving male of Woson *oobi* and, at Mome Hafi, chose to garden and fish in an area that he asserted had, at a much earlier time, been 'given' to members of his *oobi* by the previous owners (Dwyer and Minnegal 2007). During preparations for a feast, the previous owners had granted rights to fell sago palms and incubate sago grubs to Woson co-residents—their land was distant, in the lower mountains and had few palms—though Noah now extrapolated this act of generosity as being a 'gift' in perpetuity of the land and the streams in the vicinity of those sago palms. He went further, as well, by attempting to exclude other Mome Hafi residents from the area of land in question, and it was this that caused much friction. His claim was constructed as one of exclusive ownership and was demonstrated, in his view, by his current use of the land which was in keeping with conventional Kubo practice, by invoking his own version of the past and by drawing on an emerging ideology of ownership that was grounded in specification of bounded areas of land associated with particular groups of people. His claim drew upon both relational and categorical understandings of the world. To this time there were no such disputes that were not partially embedded in a demonstration of ongoing use of the land by those who were laying claim to it. Claims to ownership were not yet abstracted from practice.

Waiting

Through the years 1986 to 1999, people lived comfortably at Gwaimasi and, later, at Mome Hafi. They had little access to money or the things that money could buy. Food was usually easy to come by and their proximity to backswamps meant that sago palms were numerous. A major drought through 1997 had few adverse consequences for them, though it had seen some people join their community to escape deteriorating conditions at higher altitudes (Minnegal and Dwyer 2000a). Disputes were few, far between and usually quickly resolved. Past concerns about cannibalistic raiding parties had dissipated, though the threat of sorcery remained ever-present. Through those years, however, the people changed. In many domains of their lives, both material and social, they de-emphasised a relational understanding of the world in favour of a categorical understanding. They extrapolated the logic of money—reification, commensurability, categorisation and anonymisation—to expressions of exchange, to gender relations and to notions of use rights. They did these things themselves. The changes we witnessed were expressions of their agency, of their attempts to draw the outside world in and, thereby, gain access to what they perceived as its future possibilities for their own lives.[20]

But, through all this time, they waited. They waited to be visited by friends and kin from other communities, for when that occurred life and affinity could be celebrated with talk, laughter and feasts. They waited for timber or mining companies to come and provide what was needed to facilitate access to the rumoured world that lay beyond. They waited for God and for the return of Jesus, for here there was the promise of a different kind of salvation that would continue long after death. And, at times when we departed, they waited for our return or for the arrival of people like us who we sent because, though we had less to offer than companies or God, we had been at least a tangible presence in their lives.

Little of what they waited for eventuated. Kin and friends continued to visit, but neither God nor Jesus put in an appearance. Even we let them down because, after 1999, we never returned to the Strickland River. Nor did we send replacements. And other expected opportunities came to naught. Men came from Kiunga in the west promising a timber industry. They solicited money, departed and were never heard of again. Men came from the north, asserting that the mine at Porgera in the Highlands was polluting the river and contaminating the land. They too solicited money to 'fight' the mining company but they too departed and were never heard of again. Throughout these years the people lived comfortably but with deep anxiety about what might be, about what others had and they did not, about their future. The dead fish on the bank of the Strickland River was a sign that their anxiety was fully justified.

Through all those years of our engagement with people who lived close to the Strickland River we were impressed by the changes that seemed to be endlessly underway. But our engagement, though it spanned 13 years,

20   Throughout this book we discuss consequences for the people of their engagement with the logic of money and of their increasing desire for more money. To them, money provided opportunities to access desired goods and a lifestyle that was judged to be 'better' than that which they currently experienced (Minnegal and Dwyer 2007). They imagined that money was empowering. Bill Maurer (2006) makes many similar points in his review of the anthropology of money.

was a mere snapshot, a moment in time. There had always been change and there always would be change. Agents of the Australian colonial government, missionaries, and men seeking gold or oil, had engaged with Kubo people and their neighbours for more than two decades before we arrived. Parties of explorers had intruded on their territory at earlier times. And, more recently, officers of the National Government and the bureaucratic procedures within which they are embedded, and which they impose on others, have encompassed the lives of the people who we have come to know. These encounters are the themes of the next chapter.

3. Timelines

The social landscape of Kubo has never been static. There has always been change as people were born and died, created new alliances or dissolved old ones. But the changes observed at Gwaimasi were of a different order, grounded in an increasing awareness among the people who lived there that they were on the periphery of a much larger world, with sources of power and wealth located elsewhere. A newfound awareness of themselves as Kubo—occupying and defined by a shared place in this new world order—was tempered, too, by a realisation that new opportunities might not be distributed equally.

In this chapter we trace local histories of engagement with the wider world that shaped those changing perceptions. We begin with stories told by the first representatives of that world to reach this area, emissaries of empire and enlightenment who came seeking to know, and thus control, this 'last unknown'. In drawing up maps and censuses, pinning people to communities and places in the process, these explorers and colonial officers constructed Kubo as a distinct category of people. But the people who came to bear that name watched too, and told their own stories of what was happening—stories of relationships and interactions with each other, with the land and with the new outsiders. These we trace next, as they reveal the shifting perceptions, hopes and desires that shaped the lives of the individuals and groups we first met at Gwaimasi. As Papua New Guinea (PNG) moved towards Independence, the stories of engagement took on a somewhat different complexion. Missionaries came in search of souls to save and others came in search of resources—gold, oil and gas—to extract, each bringing with them new understandings, new possibilities and new expectations. And as resources were found near this area and plans for extraction projects took more concrete form, the state began to play a more active role in the imaginings of local people. The Department of Petroleum and Energy (DPE), in particular, through its little-understood but clearly crucial bureaucratic imperatives, has come to exert a defining role in the ways local landscapes are being reconfigured in the early twenty-first century. Its language of tribes, clans and subclans, of Incorporated Land Groups (ILGs) and Landowner Companies, of equity and royalties and business development, have increasingly infused the ways Kubo people frame identities and the strategies they devise to achieve desired ends.

The various histories we trace here interweave. In each of the four sections of this chapter we tease apart and follow particular threads through the events and changes of the past, into the 'present'—the early years of the second decade of the twenty-first century—of which we write in the rest of this book.

Stories from the Outside

We saw a catamaran-like raft on a creek; and just beyond this was a clearing in the forest, on which was a square native home. There were many birds about in the scrub. Coming round the next point, we saw several short, chisel-shaped dugouts, quite different to anything we had seen before tied up to stakes in the mud. On this rocky promontory we saw a rather light coloured, short, thickset native, with a mat of hair hanging down his back: he had a stick across his shoulder, from which hung a string of what looked like several small fish. He gazed at us for a moment, raced up the bank, and disappeared into the jungle. What must our whaleboat have looked like to this primitive man, who had never seen a white man, or anything larger than a catamaran? Our whaleboat had the sails set, three Javanese rowing on either side, the Union Jack flying above the mast, the Captain steering and the rest of us white men sitting along its sides. We lowered the sail and ran the boat up on the rocks. Shaw and I jumped ashore and crossed to where the native had been standing. We found a fishtrap set in the water between two rocks, and tied by a rattan cane to another rock. This ingenious trap was simply a large funnel of plaited rattan, lined inside with strips of the spined lawyer palm placed so that the spines curved inward: the bait was fishes in the small end of the funnel. Anything could enter the funnel trap; but, when retreating, the curved spines held the intruder on all sides. We took the trap to the whaleboat, and took some turkey red cloth, hoop iron, and a small looking glass back to the place where the native had stood (Froggatt 1936, spelling amended).

The first expedition sponsored by the Australasian Geographical Society left Sydney in June 1885 heading for the Aird River on the south coast of PNG. At Thursday Island this plan was changed. Instead, Captain Henry Everill, with 11 Europeans, 11 Javanese, a Cingalese cook and, as translators, three Kiwai men, took the *Bonito*—a 77-ton paddle steamer—up the Fly River and turned eastward into the previously unnamed and unexplored Strickland River. For two months the *Bonito* was grounded on a gravel bed more than 150 km from the mouth of the latter river. During this period of forced inactivity Everill ventured north in the whale boat. He was accompanied by five of the Europeans and six Javanese. Their most northerly campsite, on 27 September, was on the west bank of the Strickland River between the junctions of the Murray and Carrington Rivers, at the border of Kubo and Febi territories.[1]

During six days in Kubo territory, when they had respite from the arduous task of towing the whale boat upstream against the current, the explorers collected geological, botanical and zoological specimens, mapped the course of the river and observed many signs of human activity. Footprints were seen on sandy banks, trails disappeared into the forest, small houses and shelters were visible and there were rafts and canoes at various places along the river-bank. On the first day they collected two string bags and a 'fine stone hatchet' that had been abandoned only minutes before. The bags contained 'a freshly caught fish, a live crayfish, a small dead lizard, some rolled up tobacco leaf, and some ornamental bird plumes' (Froggatt 1936: 49). In exchange they left trade cloth, hoop iron and what was probably the first steel axe—a small hatchet—to reach Kubo territory. Near these gifts they placed 'a little Union Jack flag in the sand' (ibid.). But, throughout those days, they saw only one person. In 1936, 50 years after what he called 'the forgotten expedition', Walter Froggatt—twenty-seven-year-old entomologist and assistant zoologist to the expedition—drafted an account of his experiences. It was never published.

In his diary Froggatt called the rocky promontory 'Fishing Rocks'. One hundred years later, this was a favourite fishing spot of a young married woman named Babio. We ourselves fished there often and know the rocks by their Kubo name, Woimotibi. Froggatt's visit was on 25 September,

---

1   Everill (1888: 184; Anon. 1887) reported that his most northerly campsite was at lat. 5° 30' S, long. 142° 22' E, close to what was then the border of the British administered territory of Papua and the German administered territory of Kaiser Wilhelmsland. This was not correct. Dwyer et al. (2015; see also Mackay 2012) provide a revised interpretation of the travels of Everill's party on the Strickland River.

1885 and, that night, the party of explorers camped a little north of the place where Gwaimasi village would be eventually established. Three days later, returning from the north, they again landed at Fishing Rocks. The items they had left as 'payment' for the fish trap had not been touched and the canoes were still present. They found a track and followed it to a house which had a palm leaf basket hanging above the door. The basket contained a painted skull which they collected as a 'valuable ethnographic specimen'. Again, they left trade goods in exchange (Bauerlin 1886; Everill 1888; Froggatt 1936).

It was 42 years before another European passed through Kubo territory. Charles Karius arrived from the north and in spectacular fashion. On 22 May 1927, with a party of six police and 21 carriers, he boarded the first of four rafts on a 'wildly exciting' journey down the Murray River, camped overnight in Kubo territory and, the next day, continued southward by raft and canoe, reaching the coastal outpost of Daru on 10 June (Karius 1928; Craig 2013).

In the year that Walter Froggatt died and one of us was born, Jack Hides and David Lyall, with 40 Papua New Guineans, travelled through Kubo territory (Hides 1939). Hides had resigned from his position as patrol officer with the Papuan government and, in 1937, embarked on a private venture to prospect for gold (McGee 2007). He left Daru on 23 February, carried a radio to contact sponsors in Port Moresby and was irritated that country east of the Strickland had been declared an Uncontrolled Area. Officially, he was not permitted to land there.

Upstream from Lake Murray the party transferred from the 30-ton *Ronald S.* to the 32-horsepower, shallow draught *Peter Pan*. At the Rentoul River, the *Peter Pan* turned back and Hides' group continued by canoe. Late in May they passed the mouth of the Baiya (Cecilia) River and soon were within the territory of Kubo people. Progress was slow because multiple trips were needed to ferry men and supplies between campsites. It was not until 30 June that they reached the Murray River. Through this period they had several friendly encounters; at first, to the south and on the west bank, with Pare people but later, to the north and on both sides of the river, with Kubo people. Hides described the latter as 'tall, dark skinned; they wore a belly armour of bark, and their long hair hung in ringlets, matted with mud and dirt and vegetable oil' (Hides 1939: 73). He met these men on the east bank opposite the mouth of a stream that we know as Dua. They gave him 'paradise plumes, packets of red ochre, dogs' teeth necklaces, and net bags of native food' and he reciprocated with a mother-of-pearl shell and a 'bright new tomahawk'.

> The one who took both presents made a long speech, punctuating it many times by stamping his foot on the ground, and, taking in the whole of the country to the northward with a wave of his arm, repeated the word *Hakwoi*, which I took to be the name of his tribe (Hides 1939: 73–4).

The man had said '*hugwa* (come)'. He was inviting the foreigners to a house not far to the north. They did not understand, returned to their camp and, the next day, again joined people on the east bank. They talked and smoked with them and exchanged a second tomahawk for a large and fat domestic pig. They slept one more night at the Dua River campsite and then continued to the north. They were watched by men, women and children who 'waved the pearl shell and chopped the trees with the tomahawks we had given them' (Hides 1939: 76). By 1937, at least, Kubo were accepting and appreciative of steel tools.[2]

By early July, Hides' party had moved north out of Kubo territory and into, and beyond, the land of Febi-speakers. It was 1 September before they returned. They had had a hellish time. From the top of the ranges Hides had seen 'a great valley system of fertile river flats, of timbered and grass plateaux, and smooth rounded domes and slopes'. He had seen 'columns of smoke rising far up in the valley' and pictured 'the inhabitants stirring their fires to life to roast their early morning meal of taro and potato' (Hides 1939: 120). But he could go no further. Lyall was sick and nearly blind. They waited in vain for an airdrop of food and medicine, and finally their radio failed. For two weeks they carried Lyall back from the mountain top, negotiating their way south through precipitous limestone. More than 20 other men became desperately ill with beriberi; five died. When the debilitated party reached the Strickland those who were well enough spent three days building two large rafts. It rained continually. Their only sustenance was tea and saccharine tablets. The rafts were large

---

2   Froggatt (1936) reported that the man seen at Fishing Rocks was wearing a bailer shell genital cover and described the house seen in this vicinity as having slung hearths. The Upper Strickland men seen, and photographed, by Hides wore bark aprons, 'belly armour', crossed strings of beads on their chests and nose plugs (Hides 1939: 118f). It was only further south on the Strickland that Hides saw men wearing bailer shells (1939: 38f). In 1986–87 the customary dress of men living at Gwaimasi was as depicted by Hides and not as depicted by Froggatt. Further, it was only in 1986–87 that people at Gwaimasi commenced making slung hearths and, in the following years, we observed this practice being taken up by people living to the east. It seems likely that, at the time of the Everill expedition, the people who lived close to the Strickland, south of the Murray River, were not, in fact, Kubo-speakers.

and heavy but insufficient for everyone. Hides, with Lyall and the sick Papua New Guineans, moved off first. The others remained to build a third raft. The river was in flood and the men had a wild ride south through the Strickland Gorge. In calmer waters, north of the Murray Junction, they stole a dugout canoe and, that night, camped beyond Kubo territory near the Rentoul River. Two weeks later they reached Daru where David Lyall died. Hides returned to Australia. He too died before his dream of payable gold at the headwaters of the Strickland was realised by discoveries at Mount Kare and Porgera (Biersack 1999; Golub 2014).

By 1986, Kubo had no recollection of these early visits by Europeans. But, though forgotten, they must have left their mark. Everill's party left trade goods. Hides' party provided steel axes and saw them being used. By the time Hides departed, Kubo understood that somewhere else there were other kinds of human beings and other kinds of material things. They had certainly learned that it was possible to exchange food that they themselves produced for some of the material items carried by outsiders. When Europeans again entered Kubo territory, their arrival was surely confirmation of rumours that were still in circulation.

In the years before and following the Second World War, outsiders appeared quite regularly in the lowland country north of Lake Murray, and bounded by the Fly River to the west and the Strickland River to the east. These were land-based expeditions, often reliant upon trading for food with local people, which established both shorter and longer-term campsites at many different locations. Some were led by government patrol officers establishing contact with little-known people. Others were by teams from the Australasian Petroleum Company (APC) and the Island Exploration Company (IEC) prospecting for oil (APC 1940). The maps produced by these latter expeditions recorded topographic features and, with varying degrees of accuracy, the names and locations of living places and language groups. Near the Strickland River, for example, Supei (Samo) people were located on the east between the Nomad and Cecilia Rivers, Daba (Kubo) people on both sides of the Strickland north of the Cecilia, and Kanai (Konai) people west of the Strickland and south of the Blucher Range.[3]

---

3   Original APC and IEC maps are not available but details from those maps are incorporated on the map accompanying Besaparis (1959) and on Mount Leonard Murray, New Guinea S-B 54-6 S600-E14200/200 AMS T401 First Edition (AMS 1), 1942 (U.S. Army Map Service). Viewed 23 August 2015 at: www.lib.utexas.edu/maps/ams/new_guinea_500k/index.html. The name 'Daba', used for people living near the Strickland River north of the Damami River, may derive from 'Dabamisi', the name of a Kubo *oobi* with land in the south of this area. On the 1942 U.S. Army Services map, Cecilia River is named Fairfish River.

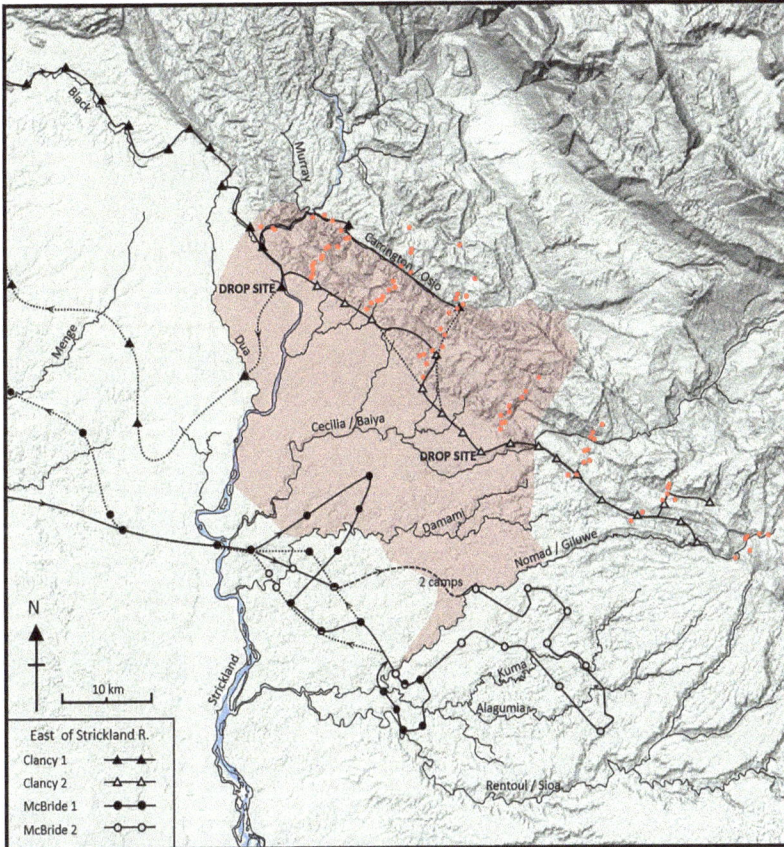

Figure 3.1: Patrols by Des Clancy and Brian McBride in, respectively, 1947–48 and 1959.

Source: Authors.

Note: Clancy's group arrived at the Strickland River on 28 December 1947 and the last members departed on 21 May 1948; McBride's group arrived at the Strickland River on 19 October 1959 and departed on 15 December 1959. Sites at which geological specimens were collected by G.A.V. Stanley's team are shown as ●. The territory of Kubo people, as reported in 2014, is highlighted.

The first official foray east of the Strickland River, and the first into Kubo territory, occurred in the early months of 1948.[4] On 3 December 1947 an APC team led by geologist George A.V. Stanley and accompanied by patrol officer Des Clancy, four other Europeans, five police and 143

---

4   Exploratory patrols by at least APC sometimes ventured, unofficially, into 'uncontrolled areas'. In 1953, for example, Dave Calder (1953) led a government patrol through southern reaches of Gebusi and Bedamuni territories and reported passing an earlier APC campsite near the junction of the Rentoul and Nomad Rivers.

'native' carriers departed from a base established on the Fly River south of Macrossan Island and headed more-or-less east (Clancy 1948; Stanley 1948; Allen 1990). From the headwaters of the Black River the party travelled southeast to reach a relatively wide stream (Dua) and then east to a tributary of the Strickland (Sigia) on 28 December (Fig. 3.1). At a Strickland River campsite they made a 45-foot canoe.

In early January, Clancy established a base camp south of Sigia, inland from the mouth of a stream named No (Dege). Here his carriers cut a drop site and for five consecutive days, from January 7, a Catalina aircraft flew low over the site and dropped supplies. Between January 12 and January 16, with 15 carriers, Clancy travelled north by canoe, turned into the Osio (Carrington) River, established a campsite well upstream, briefly met some local people and, following loss of gear when the canoe capsized, returned to base camp.

The Osio campsite was used as a base by Stanley and his assistants from 22 January to at least early April. Clancy, with police and carriers, crossed the Strickland River on 29 January and, surveying with compass and chain, travelled southeast to reach a site north of the Cecilia River and east of the present location of Suabi on 21 February. Here a second drop site—approximately 600 x 100 ft—was prepared and supplies (including freezer goods and reams of foolscap paper) were delivered on at least six days. Some of these supplies were relayed to Stanley's party on Osio River. This latter group conducted surveys south to the transect established by Clancy and in two locations up to 7 km north of Osio River (Fig. 3.1). Having established a site from which Stanley's party could be serviced, Clancy now headed further east, crossed the Cecilia and cut southeast, to reach the Nomad River. He was establishing campsites and food stores that would be used by Stanley's team as they surveyed the Cecilia anticline (Craig and Warvakai 2009). It was only at his furthest campsite, beside the Nomad River, that Clancy met relatively large groups of men, women and children. They were Bedamuni. After an initial tense encounter, bananas were exchanged for beads and steel. By 6 April, Clancy's party was heading back to the Strickland, reaching the base camp four days later.

Local people who visited the Osio campsite and, eventually, the Cecilia drop site were few in number; they were probably Kubo and, perhaps, Febi. In the vicinity of the Strickland River base camp, there were many signs of people—tracks, small gardens, sago processing sites—but, at least at first, people were wary. By mid-April it seemed that Pare people from the south and southwest were visiting the camp. Clancy left the Strickland for Kiunga on 25 April, leaving the geologists to finalise their work. The last members of the team left the area on 21 May.

Clancy had mixed feelings about the area east of the Strickland. In his diary he noted that 'cassowary and the usual pile of birds were obtained' and continued 'no tinned meat has been necessary since we arrived at this huntsman's paradise' (Clancy 1948: Diary 3 March 1948). In his concluding remarks, however, he observed that the population was 'very sparse, and it would hardly be worth the time and energy expended to patrol the area again'.

There were, however, further patrols—though it was not till 1968 that these again ventured north of the Cecilia River, where Clancy had spent most of his time. Rather, attention focused on the area between the Cecilia and Rentoul Rivers. In late 1959, Brian McBride with two Europeans, ten police and more than 50 carriers led a patrol from Kiunga southwest through Pare territory to reach a pre-established base camp west of the Strickland River and a little south of the junction with the Cecilia on 19 October (McBride 1960). In the course of the next seven weeks, McBride undertook two patrols east of the Strickland (Fig. 3.1). The first intruded into Kubo territory between the Damami and Cecilia Rivers, cut southeast through Samo territory, and followed a circular route through the heart of Gebusi territory before returning to the Strickland River camp. The second patrol traversed Samo territory from west to east, and then crossed the Nomad River to enter Bedamuni territory where, through the next week, there was a series of often tense, and sometimes frightening, encounters with relatively large groups of armed Bedamuni men.

These early encounters with Bedamuni revealed, first, that their territory was more densely populated than land to the west and northwest where Kubo, Samo and Gebusi people lived and, secondly, suggested that Bedamuni, more than their neighbours, were engaged in frequent intertribal and intratribal fighting, cannibalism and undesirable mortuary practices. For these reasons, R. Lang's extended patrol of 1961 was charged with establishing a government station on the Nomad River (Lang 1961a, 1961b, 1961c, 1962a, 1962b; Russell 1962; Stott 1962).

From here it was thought that the 'fierce' Bedamuni could be brought under control and the supposedly less aggressive Kubo, Samo and Gebusi guided towards more 'civilised' ways of living.

It took 18 months for Lang's men to build an airstrip and associated facilities at Nomad. They themselves received supplies that were carried overland and ferried across the Strickland near its junction with Baiya. A labour force of more than 150 men, mostly outsiders recruited from Kiunga, Debepare and Lake Murray, was employed. Some people living within a day's walk of Nomad were employed as labourers, encouraged to adopt a more sedentary lifestyle and ordered to abandon customary burial practices and cannibalistic raiding. They gained access to European clothing and steel tools—a steel axe-head in payment for a month's work—but it would be some time before they were introduced to the anonymity that characterises monetary transactions.

From 1963 onward, with Nomad Station fully operational, there was a more concerted effort to contact outlying communities. Near at hand, rest houses for European patrol officers and barracks for police were built. Some of these attracted larger communities of Samo and southeastern Kubo people (Suda 1990; Shaw 1996: 34–7). To the east, in Bedamuni territory, a regular and well-armed government presence was necessary before raiding was reduced. In 1964, for example, Bedamuni people burned a Kubo longhouse to the ground, killed four people and kidnapped three girls. Later, in the same year, matches, soap, mirrors, knives, 24 axes and three Australian flags were stolen during a raid on the equipment store at Nomad. In 1967 Kubo people killed four Bedamuni and carried two bodies away to eat (Anderson 1970: 19; Barclay 2012). On these occasions and others, men were arrested and, after being tried, often jailed at the distant coastal town of Daru 'partly for security reasons and partly to broaden the horizons of these isolated bush peoples' (Hoad 1964).

Kubo people who lived north of the Baiya were seldom visited. There were only a few hundred of them living in widely dispersed communities. Tracks in the area were difficult to follow and the rivers were often in flood. Local guides were uncooperative and, most importantly, from the viewpoint of administrators, the people were not troublesome. They were easy to ignore. Indeed, for much of this time, they were regarded as a separate language group, more closely related to southeastern Kubo than to Samo but bracketed with Febi of the mountains to the north as the Daba.

Through a five-year period from 1964 to 1969, patrol officers Bob Hoad, Allan Johnson and William Patterson spent a total of 28 days north of the Baiya (Hoad 1964; Johnson 1968; Patterson 1969a, 1969b). These patrols, however, were on the western side of Kubo and Febi territories close to the Strickland River. At most they met only 15 to 20 people at each community they visited. They reported that wild pigs and sago were readily available and that the supply of garden food was less reliable. They mistakenly thought there was little interaction between neighbouring communities, and where, by chance, they met small groups temporarily based at a bush house—a base for hunting or processing sago—were predisposed to identifying the people as 'true' nomads. They reported the relatively slow penetration of European artefacts into this area and the people's enthusiasm for steel axes and machetes.

In 1969, Patterson's group spent several days at Sigiafoihau, on the west bank of the Strickland. They purchased food, processed a sago palm, provided medical treatment and found the people hospitable. They did not know that, either shortly before or shortly after their visit, these people hosted a party at which five of the guests were killed and eaten. In 1987 we snacked on breadfruit at the long-abandoned site of that longhouse, only 20 minutes north of Gwaimasi. The old clearing was now overgrown with bracken fern, and the forest was closing in. Where the ground had been turned by a pig, we found part of a stone axe. Uhabo and Geswa had cooked the breadfruit while their wives, Fafobia and Tiotidua, processed sago nearby. The two men were of Gumososo *oobi* and this was their land. When Patterson left Sigiafoihau, Uhabo guided them up Sigia stream into the territory of the Konai-speaking Watia clan. He was one of the guides that Patterson reported 'was of much more use in showing us the way out of an area than into it'. Repercussions of that cannibal feast, of the tensions it generated, of initiations held and marriages arranged as attempts to restore good relations, continue to the present day.[5]

From May to August 1970, W.A. Cawthorn led a major patrol from Nomad that travelled north to the Baiya River, northeast to the Osio River, west to the Masi River, north through the territory of Febi people of Wuo *oobi* to the southern reaches of Bogaia territory, south along the

---

5   In May 1969, ten days after Patterson's visit, John McGregor (1969), patrolling from Kiunga, listed the names of 16 people from at least three Kubo *oobi* and two Febi *oobi* at Sigiafoihau. Seven of the people named, including all five children, were alive in August 1986. All were then living at Gwaimasi, about 15 minutes walk from Sigiafoihau.

Strickland River and via communities living in the wedge of land between the Strickland and Murray Rivers to Magwibi, between the Baiya and Damami Rivers. From Magwibi the patrol moved east to the Damami and, after six days, reached Adumari in Bedamuni territory (Cawthorn 1970). This was the first patrol since Des Clancy's visit in 1948 to traverse central Kubo territory and eastern Febi territory, and only the second to meet people living in the southern reaches of the Strickland-Murray wedge.[6] No people were encountered during either the traverse of central Kubo territory or the journey that followed the Damami River into Bedamuni territory though scattered groups, including people of seven *oobi*, were met, or spoken of, in Febi territory. In the early 1960s, raiding by Bedamuni had led Kubo people living in the eastern portion of their territory, north and south of the Baiya River, to join communities to the west.[7] Indeed, people at Magwibi told Cawthorn that the land to their east, along the Damami, was occupied by Bedamuni. It was not until after this patrol, and in response to both the increasing pacification of Bedamuni and the year-long presence of the petroleum company Texaco in the upper reaches of the Baiya River, that Kubo people resettled these portions of their territory (Meintjes 1972, 1973).

Cawthorn had intended to lead his patrol into the headwaters of the Damami and Baiya Rivers on the western slopes of the Karius Range. It was known that some people lived here, but they had not yet been contacted by government patrols. Time constraints and a shortage of food meant that he was unable to accomplish this aim. However, Rob Barclay led patrols into this area in 1971 and 1972 and contacted some small groups of people—variously named in patrol reports as Siali, Sialu and Siae—who he considered to be more closely aligned by language and clothing to Etoro, on the southern slopes of Mount Sisa, than to either Bedamuni or Kubo (Barclay 1971a, 1971b, 1972).[8] They lived at moderately high altitudes—750 metres above sea level and higher—and appeared to be more reliant on tubers, including sweet potato, than sago

---

6   In 1968 the anthropologist Frederik Barth visited communities in the Strickland-Murray wedge when he accompanied John McGregor on a patrol from Olsobip (Barth 1971).

7   Our understanding that Kubo people had effectively abandoned the northeastern reaches of their territory through the early 1960s, and did so in response to Bedamuni raids, is based on both interpretations of patrol reports and some accounts by Kubo people themselves. In recent oral accounts it is sometimes suggested that the Bedamuni raiders were acting as 'mercenaries' on behalf of eastern Kubo people.

8   It is probable that descendants of the people named as Siali, Sialu or Siae in the early 1970s now reside at Gesesu in Febi territory, self-identify as Tsiani (or Sabalimatie Tsiani) and are referred to by Febi, Kubo and Huli as Mora (Denham et al. 2009: 4.23, 4.60–66).

and more reliant on hunting and gathering than on gardening. In 1972, Barclay undertook a lengthy patrol that covered the western slopes of the Karius Range and, in Febi territory, revisited eastern, northern and western areas that, earlier, had been patrolled by Cawthorn.

Barclay (1971a) and others refer to the devastating effects of three influenza epidemics that swept through populations near Nomad in the late 1960s and early 1970s. The source of these epidemics was attributed to employees of the companies Texaco and Compagnie Generale Geophysique, exploring for oil and gas in the region. Among Bedamuni alone a minimum of 168 influenza-related deaths were recorded—estimated at more than 5 per cent of the population—but there are no estimates from other language groups in the area. Barclay commented that deaths from influenza were usually taken to implicate sorcery and sometimes resulted in sorcery killings which continued to and beyond this time. He noted that most of those implicated in these killings were males between 18 and 25 years old. He suggested that, in response to recent changes in their society:

> there is a desire by youths to 'prove' themselves to be more mature members of the community. Hence, when dying relatives indicate the sorcerer (or the sorcerer is discovered by other means), it is invariably the young men who volunteer as executioners.

Throughout the 1960s, northern Kubo communities and their Konai and Febi neighbours were viewed as a remote backwater. Hoad judged them to be among 'the most primitive people I have encountered', often living completely 'within the stone age' with 'little trade, little wealth and little promise for the future. To them their horizon has no allure' (Hoad 1964). In the mid-1960s they seemed timid, avoiding Hoad's party of more than 50 carriers, but by the close of that decade were much more welcoming of outsiders. They were relatively healthy, contented and, though ignorant of new laws, not disruptive. In Hoad's words they were 'incomparably lovely'.

To those early patrol officers, the decade of the 1960s offered risk, adventure and new experiences. It was an exciting period of their lives. But it could not be sustained. By the early 1970s Nomad Patrol Post was well established, the 'fierce' Bedamuni were settling down, and for a new generation of patrol officers, some with wives and young children, life was becoming more mundane. They continued to visit outlying communities, but these were now known; there was less that was new.

The tasks became routine: taking a census of people who did not come when wanted, reprimanding them because trails and village houses had not been maintained, attending to festering wounds and other health needs, and acting to minimise and punish theft of government property. By October 1973, most Kubo people, together with some Konai and Febi, were aligned with six communities.[9] These assemblages were short-lived, however. Through the mid-1970s, many northern Kubo people dispersed as smaller longhouse communities and many Konai and Febi moved back to their own lands.

Amalgamation as larger communities and its possible consequences were seen in different lights by different patrol officers. Cawthorn (1970) argued that:

> it is a moot point whether village grouping at this stage is a good thing … socially it would have a disruptive effect on their society if implemented too early … to control nomadic groups a government must first change their nomadic way of life. But, in doing this it must be careful not to alienate them by destroying their traditional society (Cawthorn 1970: 27).

Others felt more frustrated at what they saw as the recalcitrance of the northern Kubo. Philip Fitzpatrick's (1971) guides were uncooperative and, at one village, the performance of the probationary village constable had been unsatisfactory. He wrote:

> The average Supei/Kubor, in his relatively idyllic albeit apathetic situation does not particularly wish to become politically aware, in fact, given his daily bread and license to do as he please, he doesn't care at all about what is happening in his country. The political situation in the Upper Strickland Census Division could best be described as 'stagnant' (Fitzpatrick 1971: Situation Report 3).

---

9   Census records from late 1973 record Sigiafoihaimuson west of the Strickland and south of the Murray River junction (near the future location of Gwaimasi) with an estimated population of 43, Headubi east of the Strickland and north of the Baiya with 140, Soabi 2 which was close to Wa stream and east of the former Soabi with 71, Magwibi and Udamobi east of the Strickland but south of the Boye with, respectively, 143 and 85, and Siuhamuson (later Testabi) north of Nomad with 85 people (Meintjes 1973). Sigiafoihaimuson was probably mixed Kubo, Konai and Febi while Udamobi and Siuhamuson were probably mixed Kubo and Samo. Many of the recorded people probably lived elsewhere but, having advance knowledge that a census would be taken, satisfied government requirements by assembling at a focal village.

But the people did care. They were vitally interested in what was happening on and near their land. Change was rapid; there was little consistency from year to year. Through the 1970s missionaries came, companies exploring for gold and oil continued to visit the area, an experimental agricultural station was established at Nomad, linguists and anthropologists lived among Samo and Bedamuni people. PNG became self-governing on 1 December 1973 and an independent nation on 16 September 1975. The last expatriate patrol officer left Nomad in early 1976 and, thereafter, the station was run by trained Papua New Guineans. These men were threatened with, and anxious about, sorcery; patrols through regions, and among people, that were still poorly known rapidly declined (Robin Barclay and Laurie Meintjes, personal communications, 25–26 November 2014; Mike Milne, personal communication, 1 January 2015). From 1974 to beyond the year 2000 there were no visits by government officers to Strickland River communities in areas to the immediate north and south of the Murray River junction.

Stories from the Inside

When APC [Australasian Petroleum Company] came to the Strickland River they gave a 'white' woman to the older brother of Asekai's father. He was a young man at that time, and not yet married. The woman stayed with him and they were happy together. But when the white men came back there was a problem. The white woman's husband was quarrelling with another man about ownership of a bark cloth. The white men misunderstood. They thought the quarrel was over the woman. They said that if she was going to cause trouble then they would take her away. And that is what they did. Her husband was very sad.

The APC camp was made at Dege Hafi—the place where the stream Dege joins the Strickland River. No is another name for Dege. The old men did not behave well—they stole things—and the white men left. If the old men had been better behaved then the white men would have stayed. They would have built 'Mosebi' at Dege Hafi but they built it at Nomad instead.

These brief stories date from 1948, when APC visited Kubo territory. On that visit, a base camp was established at the stream named No. Memories of that camp prompt the earliest stories that Kubo tell of the arrival of white people. The second is mirrored in the east of Kubo territory, where it was another 1948 APC base camp, this one at Auma Hafi near the Baiya River, that people say might have become 'Mosebi'.

In the mid-1980s Kubo knew little of Port Moresby. To them, Mosebi was the centre from which development would come. It held almost magnetic attraction. In 1995, people at Gwaimasi discussed moving the site of their village. Should they move one hour's walk to the northwest to a site on the bank of Sigia stream or one hour's walk to the south to a site on the bank of Mome? They chose the latter. Mome, they argued, was in the direction of Mosebi. Sigia took them in the wrong direction both in space and in time: 'we would be going backwards', they said.

When white men first visited Kubo territory they were named *bou*. They were spirit beings. But *bou* assumed a more general referent and came to connote outsiders in positions of authority: patrol officers, police, company employees, missionaries and anthropologists, among others, irrespective of the colour of their skin. As the years passed by, and the English language took hold, it was sometimes necessary to back-translate. In some contexts *bou* became 'white' and the woman who may or may not have lived for a time with the older brother of Asekai's father may well have had brown skin. There is, of course, no record of carriers, let alone white men, taking female companions with them on early ventures exploring for gold or oil. That might have been frowned upon.[10]

In our earlier years with Kubo they told entertaining stories of the trickster, Dikima, and his fall guy, the buffoon Wamagosai. In one tale, for example, Dikima chews on a vine, the sap runs and his teeth become black. Wamagosai is jealous; he too wants black teeth. Following Dikima's untruthful instructions, he climbs a tree, tells his grandmother to light a fire underneath and, overcome by smoke, crashes to the ground. When children hear a tree falling in the distance they may be told that it is really Wamagosai. Vanity, they may come to understand, can have unwanted consequences.

In other stories an animal—a fish, frog or flying fox, for example—appears in the form of a man or a woman and lures a partner to an alternative world, below the surface of the water or in the forest canopy,

---

10   Five wives to policemen accompanied the James Taylor and John Black Hagen-Sepik patrol of 1938–39, though Taylor preferred that this not be generally known (Gammage 1998: 2, 209–10).

requiring them to behave in certain ways. If the partner behaves according to instructions then marriage and a family might follow; if not then the human-animal departs leaving the prospective partner alone and sad. One such story concerns the mullet woman.

A man lay still, with his eyes closed, pretending to be dead. People painted him with red ochre and rubbed his body with juices from a dead and rotting bandicoot. They carried him to a stony beach and they cried. The man opened his eyes a little and watched them mourning. Men and women were present but after several days and nights all but one big woman departed. The man opened his eyes, stood up and took the woman to his house. He told her that she should sleep on the woman's side of the house and he would sleep on the men's side. That is what they did but, at night, the man would wake. He could hear a noise like that made by a fish when it flapped its tail on land. He thought he must be hearing a fish in the distance. One day, after they had been together for some time, an old man came. He had much skin fungus. He invited them to a feast but said that first they must kill wild pigs and dry the meat and prepare sago grubs. They did these things but at night when the man ate some of the pig the woman ate only worms and small black snakes. When it was time for the feast they carried their offerings and walked to the river. The woman told the man to hold her hand and close his eyes. He did so and when she said he could open his eyes again he saw that they had come to a large house with many men and women present. They were given sago to eat and tobacco smoke. There was dancing and singing and they watched this until morning. The old man returned. He told them that other men were coming to kill them and that they should be ready with their bows and arrows. There was a big fight. Everyone was killed except the man and the big woman. These two started to return to their own house. They reached the river and the woman again told the man to hold her hand and close his eyes. He did so and when he opened his eyes they were at their own house. Now, however, a second house had been built and people were present eating fish. The woman told the man that he could visit that house but he must not eat fish. 'If you eat fish I will not stay with you.' She went to her own house. The man went to the other house, saw that everyone was eating fish, was tempted and ate fish too. When he returned to his own house the woman smelled the fish. She got her string bag and went to the nearby stream. She jumped in. She cut through her legs in such a way that she was now a fish. She was a mullet—a Tio—with markings like skin fungus.

Occasionally we elicited sketchy origin tales: the man who poured water from a cylinder of bamboo to create the Strickland River and caused the land of Headubi *oobi* to be divided. There were 'just so' stories that provided prompts about food taboos. There was a time, for example, when the crocodile lived in trees and the water goanna was unable to climb. The crocodile was heavy and often fell to the ground. The two creatures changed places and that is how they live today. Each is ancestral to one of two neighbouring *oobi* and each was taboo to its descendants. Or, again, the white and black spotted cuscus Baiamo once rested in a hole in the ground and the grey cuscus Nawi rested in trees. But Baiamo was easy to see by men who looked into ground holes. So the two kinds of cuscus changed places and now, when men look up into trees, they do not detect Baiamo because the broken pattern of its markings looks like leaves against the sky. There were, as well, stories of cannibalism, always with the caveat that 'we've stopped eating people now', about the murderous raiding and the burning of longhouses by Bedamuni men and, always with feeling, stories of deaths due to sorcery or, more hesitantly, hints about the killing of sorcerers.

A few stories explained connections between distant but related *oobi* (Dwyer et al. 1993; Minnegal and Dwyer 2011a). Headubi is the name of a constellation of at least six *oobi* holding non-contiguous areas of land within Kubo, Febi and Konai territories. The originating group occupies land northwest of Suabi, spanning the Strickland River. Dogs there fought over a bitch in heat. One dog, chased by others, fled into a cave. It emerged 25 km to the southeast on a small island—Sodiboko—in Damami River south of Suabi. The descendants of that dog are the Damami River Headubi. Another dog emerged far to the north in the land of Febi people, from a hole that can still be seen, and gave rise to the Febi line of Headubi. Sisiti Nomo tell of a pig that entered a cave in their land, beside the Baiya River near where Suabi is now located, and emerged from a hole near the centre of Bedamuni land; men of the Bedamuni clan Kebo (pig) are recognised as their brothers. Gumososo and Dumiti are 'brother' *oobi*, unable to intermarry though their lands are not contiguous, because an underground road once linked those lands until a falling tree blocked the entrance at Doitafa despite the best efforts of a bird to hold it up. These are not stories of migration, like those characteristic of Highlands societies—stories that trace journeys made overland by ancestors, recording interactions with people and places along routes and thus providing a charter of claims to land along the way. Only one story came close to that form. The Kubo *oobi* named Yawuasoso and a Febi clan of the same name recount how, long ago, one man travelled north and another south, swapping places and carrying with them plants and animals that otherwise would have been found only in their original setting. But this, too, is a story of connection between social groups, not with land. It has little to say of the land traversed on the respective journeys of these two ancestors but, rather, establishes a 'brother' relationship between men of different *oobi* and, through this, a prohibition on intermarriage.

In the early years of our research, however, the deep past was not focal for Kubo (compare Bird-David 2004). There was little ambiguity about access to land and they did not yet feel the need to establish rights by reporting mythological connection. That would not come for 20 years from the time we first met them. Instead, the stories that they commonly told—and continue to tell—were about travel within and beyond their own territory, about the aggregation of communities in response to government wishes, about the influence of missions, about formal education and the possibilities of employment or business opportunities. Noah's personal history contains all these elements.

> When Noah was a small boy 'enemies' came from the east, attacked the mountain community where he lived, and killed his parents. His youngest sister, a nursing baby, was killed by the same arrow that killed his mother. The enemies burned the longhouse where everyone lived. They killed pigs, destroyed gardens, and abducted Noah's teenaged sister Sesegei. Noah was about ten years old. The attack occurred in the late 1960s.

> Noah recalled his fear and confusion. 'Where is my father?' he asked a big man who was covered in black and white paint. 'He has gone to get sago,' the man lied. But Noah, too, was taken by the invaders. He lived with them for a year or so, was reclaimed by his kin, met a 'white man' who was conducting a census, travelled with him to Udamobi, south of the Baiya River, was given his first European clothing and, from time to time, attended a recently established Seventh-day Adventist (SDA) school where Tok Pisin was the language of instruction.

> In 1976, less than a year after Independence, and soon after he had been initiated, Noah moved to Nomad to continue his schooling. The school at Udamobi had closed. At Nomad he met a man from Tari. That man encouraged him to go to the Highlands and further his education. Some people said that he should not go but he ignored them. 'I decided for myself,' he said—a recurring refrain throughout Noah's account. In 1979–80 he walked more than 100 km through the lands of Bedamuni, Etoro and Huli speakers, enrolled at a school in Tari and, by 1984, had completed his primary level education. He worked as a houseboy but was often hungry.

Now, because he could not afford school fees, Noah rejoined his sister at the newly established Suabi Mission Station. She was with her third husband, a Yawuasoso man who gave Noah money to assist with further schooling. He returned to Tari but was growing restless and school was less attractive. By late 1986 he had come back to Suabi where, resisting encouragement from western kin to realign with the Evangelical Church of Papua New Guinea and live with them, he helped another man establish a small SDA community north of the Baiya River. Suabi, and the SDA Church, remained his focus for the next 13 years. He married, divorced and remarried, became father to a daughter and in 1997—the year of a major drought—relocated to the newly established community at Mome Hafi on the Strickland River.

Noah was appointed as Kaunsil (Council) for three relatively close villages. He facilitated access to an annual government grant that paid money for cutting 'roads' through forest, cleaning villages and building churches. He and his family—he now had a son as well as a daughter—followed SDA conventions of worship, though no one else at Mome Hafi did so. A dispute emerged over access to land.[11] Tensions heightened. Noah was often away, negotiating with government officials at Nomad. He felt that others gave no help to his wife and children during his absences. And Mome Hafi lacked services. There was no school or community health centre. There was no airstrip or radio. In 2006, Noah suggested that he should use some of the annual government grant—PGK1,500 in that year—to build 'Komagato Corner' at Suabi; a few houses where people from Strickland River communities could stay when they visited or children could live if they chose to attend school at Suabi. It was not to be. No one, he told us, would listen to him or do the things that needed doing. So 'I withdrew myself' from the community, he said.

Noah returned to Suabi, experimented with growing rice and wondered about the potential for a tourism business. He built an SDA church on the north side of the Baiya River and facilitated services. He was a local SDA leader. But he had not been formally trained as a pastor. He was told by church leaders to leave Suabi for two years, take up the challenge of training and, if successful, relocate to another community to preach and to guide people in how to run a church. Noah was diffident. He didn't want to go. He was getting older and wondered who would care for him in years to come. But the church leaders insisted and Noah left. It was only here, when Noah recounted his history, that he did not say 'I decided for myself'. He was pessimistic. He was approaching his mid-fifties.

---

11   See Dwyer and Minnegal (2007) for more details concerning Noah's time at Mome Hafi. Some details of his account in 2014 differ from those he provided in 1995.

His political and economic ventures had not been successes. A position in the church hierarchy was the remaining possibility. Perhaps in the future he would be posted back to Suabi, to his own people. But he knew that, by then, he would be an old man.

By 1963 Nomad Patrol Station was fully operational and, in the years that followed, people to the immediate north and east of Nomad were visited regularly to conduct censuses and, as necessary, enforce colonial laws. From that time onwards, people living in these areas—Samo, Bedamuni and southern Kubo—progressively assembled as larger and, ultimately, more sedentary communities (Shaw 1996). To the north, however, in Kubo territory across the Baiya River and in Febi territory in the mountains, such visits were much less common. For at least ten years, people in those areas continued to live as relatively well separated longhouse communities and continued to engage in both intercommunity raiding and sorcery killings.

Though seldom visited by government officers, the northerners heard much of what was happening to their south. They travelled more and attended feasts in communities that, previously, they had rarely visited. Young men had illicit love affairs. They saw new ways of living, new kinds of clothing and new material goods. They desired these things. So, where possible, some joined patrols as guides or carriers, made their way to Nomad seeking work that might be rewarded with clothing or an axe, or sought enrolment in primary school where outside languages—Tok Pisin and even English—were on offer.

As a boy, Edisa joined a patrol in the Strickland River area and travelled to Nomad to learn Tok Pisin and become a *tanim tok*—a translator. He did not enjoy Nomad, ran away and made his way back home. Roisy had more success. In the very early 1970s he too went to Nomad, to enrol in an SDA-sponsored literacy class. He did well and his teacher helped him travel to Wabag, in the Highlands, where by 1979 he had completed Grade 8. He found employment with petroleum companies, continued schooling with the College of External Studies and, in 1982, earlier than any other Kubo person, had completed courses to the 'Form IV level'. It was 13 years before Roisy came back to Suabi. He had received word that his brother had died. Even now, however, he continued to work with petroleum companies in the Strickland River area. He did so to the mid-1990s, when he married and redirected attention to raising a family and training as an SDA pastor.

Other young men were also on the move. As a newly initiated, young unmarried man Digimo moved from Kesomo land at the headwaters of Toio stream to a longhouse community near the river Wa, about 5 km from the future site of Suabi. This was before Independence. He came, his widow told us, because 'government' had said that all the 'bush people' who lived scattered across the land should come together at one place. Years later, at Suabi, Digimo assumed the role of pastor to his own people. He died in 2005. Sisiagwei, of Bosua clan, moved from the mountains at much the same time and for the same reason. He joined a longhouse community near Baiya River, in western Kubo land. He encouraged his sister to join him but she refused. She stayed behind to look after her pig, Biyohwo—the 'child' of a Bosua mountain named Biyo—and that is where she died. Atimu came from the mountains a little later. He too, like Digimo, moved to the community near Wa. But the movements of these young men were not entirely unprompted. At more southerly communities patrol officers had encouraged older men to bring in the 'bush people'. To the west, Modu took on this role; to the east it was Habukau.

A few youths and young men, like Noah and Roisy, pursued the option of formal education, travelled to Nomad—often under the auspices of SDA—and, in some case, moved further afield. They were the exceptions. More youths and young men drew nearer to the government centre. But they were cautious. Raiding had eased but sorcery accusations and sorcery killings continued. Many of those who moved south from the mountains remained within Kubo territory, where at the least they had affinal links, perhaps even in-married sisters. By the mid-1970s many people had moved to newly established villages north and south of the Baiya River. Subsequently, however, these communities fragmented as people returned to their own lands and, in the case of Kubo, recolonised lands close to, and along the length of, Baiya River. When, in the early 1980s, the airstrip at Suabi was initiated there was a handy workforce living nearby at Wa River. The people there were from multiple clans. The site selected for the airstrip lay where the lands of three *oobi*—Andibi, Domiti and Baiyameti—converged. These three, with Gobogometi, Osomei, Sisiti and Wamiti, 'came together' to build Suabi, and they continue to be credited with bringing the associated benefits to the community. The airstrip project, as it neared completion, attracted more mountain people, both Kubo and Febi. By the mid-1980s there were no Kubo longhouse communities at higher altitudes, though people still visited their clan lands to hunt and fish and make gardens. Through the early 1980s, Febi people established their own permanent 'corner' at Suabi though, in fact, some Febi had lived with Kubo affines for so long that they had become, effectively, Kubo. The movement of Febi people to Suabi has continued since that time. A dispute in 1995 saw a group from Siabi relocate to Suabi and, more recently, a perceived need to have access to the outside world has seen more people arrive. The airstrip also attracted as many as 50 Bedamuni who settled on the southern side of Baiya River but, by the early 1990s, in response to emerging tensions, they had returned to their own land.

Now, in the second decade of the twenty-first century, 50 years after the Nomad Patrol Station was opened, the stories young men and some young women tell of themselves, or the stories that are told about them, are, at one level, different and, at another level, the same. The magnetic pull of Nomad has given way to the attraction of Kiunga in the west, the Highlands to the northeast and Port Moresby. People are still exploring the world beyond. As before, they are still both excited and anxious. As before, some who go will remain away for decades or, perhaps, never return—though now, since 2011 they can sometimes communicate by mobile telephone. Young men and young women are doing what their parents did before them. But the scale is different and the consequences for life at home—for the lives of those left behind—will also be different.

Seeking Souls, Seeking Resources

In 1963, one year after the Nomad Patrol Station was established, the Asia Pacific Christian Mission (APCM)—later named the Evangelical Church of Papua New Guinea (ECPNG)—built an airstrip at Honinabi within Samo territory. The small mission station was staffed by two Papua New Guineans—a pastor and medical orderly—who ran a small trade store and initiated instruction in Christian beliefs and the English language (Shaw 1990). In the late 1960s a remarkable Australian, Tom Hoey, established a mission at Mougulu within Bedamuni territory. He and his growing family lived there for more than 20 years. He built an airstrip, wire suspension bridges across major streams and a small hydro-electric plant, and he facilitated the development of schools and community health centres (Dwyer and Minnegal 2007: 259). He continues to visit. His influence is felt throughout the region.

Dan Shaw, an American missionary-linguist-anthropologist affiliated with both the APCM and the Summer Institute of Linguistics, settled in Samo territory in 1970 and, a year later, Papua New Guinean pastors of the SDA Church opened a few schools at Nomad and further north towards the Damami River (Shaw 1990, 1996), including the school at Udamobi that Noah attended. By the late 1970s, SDA was established within southeastern Kubo at a community now known as Testabi, and had moved north across the Damami where, unsuccessfully, they attempted to build an airstrip at Tiamobi.

In the late 1970s an American missionary, John Fletcher, came to Honinabi. He sought a suitable site for an airstrip and mission station north of the Baiya River in the east of Kubo territory. His first choice was vetoed by the experienced Tom Hoey. The final choice was an area of unused forest that, to local people, was a sacred resting place—a *toi sa*—of the spirits of the dead. The new mission station was named Suabi, the third location in official records to carry this name. A Bedamuni man, Fiagone, had preceded John's arrival. He was the first pastor here.[12] Later, when the airstrip was completed, John relocated to Suabi from Honinabi. In the early 1990s John and his wife, Celia, returned to the United States of America and were replaced by Tom and Vicki Covington who, with their growing family, stayed for more than ten years.

To the west, on the border of Konai and Kubo territories, in 1979 a Papua New Guinean ECPNG pastor brought 60 or so Konai, Kubo and Febi people together at Sesanabi, on a tributary of the Strickland River. Pastor Krubi was replaced by Pastor Mama in 1982, and in 1985, soon after the latter departed, the community fragmented (Minnegal and Dwyer 2000b: 44). By the late 1980s, the Christian Brethren Church had established a foothold at Tobi in the land of Febi people.

The ECPNG missionaries and their wives were talented and practical people. They learned local languages, translated the Bible, taught pastors, promoted literacy, maintained airstrips, went on patrol to distant communities, provided medical and obstetric care, initiated and assisted local business ventures, facilitated the work of anthropologists and others and patiently cajoled government authorities into providing schools, community health centres and employment opportunities. But always

---

12   In 2014 people at Suabi celebrated Fiagone's contribution to their lives at a market day event with speeches, reminiscences and a substantial offering of food, material goods and money.

they, and particularly their sponsors living in distant places, understood their primary purpose to be more serious. By 'planting' the Word of God they were shedding light where there had been darkness; they were freeing local people from the evil influence of Satan. Where possible, and within the limit of their abilities, they strove to make their own teaching culturally appropriate. As Dan Shaw expressed it: 'As an anthropologist and Bible translator I was interested in combining the two disciplines … in order to make the "message" understandable but minimally disruptive to a specific group of people' (1990: 6).

Shaw was concerned by the legalistic emphasis of the Christian instruction preached by many Papua New Guinean pastors: 'don't smoke, chew betelnut, eat people, or fornicate' (Shaw 1990: 182). He argued that this alienated people from the work of the mission. Indeed, it was the pastors, rather than the European missionaries themselves, who, with varying degrees of success, discouraged local birth control practices and forms of ritual and ceremonial expression. And it was the former, though with the compliance of the latter, who routinely promoted the long-standing wish of government that people should live as larger and more permanent communities. As much as people tried to oblige, their customary ways of living were disrupted and they became increasingly frustrated when expectations, and the imagined promises of others, were not fulfilled.

While the missionaries came to find untouched people and 'give' to them, others came to find untouched resources and to 'take'. They too took advantage of the airstrip at Nomad. In about 1968 an oil exploration company, Compagnie Generale Geophysique, visited little-known areas to the north of Nomad and in 1969–70, for 12 months, Texaco undertook exploration work and drilling in the headwaters of the Baiya River. In 1982 Gulf Oil Corporation initiated geo-surveys and drilled for oil at Juha, in mountains well to the north of Nomad (Goldman and Ernst 2008: 9) though, at this time, their access point was Tari.[13] In 1984 they completed the Suabi airstrip to facilitate access to Juha. Additional geo-surveys and seismic surveys were conducted in the area in 1991 and 1994 (Goldman and Ernst 2008: 9). To the northwest, across the Strickland River, other companies undertook exploratory work in 1985

---

13   Hides (1939: 95) reported that 'Juha' was the local name for the Strickland River in the area near its confluence with the Murray. It is likely that his rendering was a mishearing of the name we know—Wua.

at the Kubo–Konai border and, between April and July 1989, from a base at the Konai community of Dahamo, conducted seismic and geo-surveys in foothills and mountains to the east.

None of these early ventures led to extraction. The promising Juha sites had been put on hold and it was not until the mid-2000s that they were revisited and two more wells drilled. At that time Esso Highlands, a subsidiary of ExxonMobil, was the primary operator. Through much of 2006 and 2007 they used Suabi as a base. Eventually, however, a conglomerate of companies joined forces to form the integrated Papua New Guinea Liquefied Natural Gas (PNG LNG) Project: the project was led by ExxonMobil and participating partners were Oil Search, Santos, Nippon Oil, National Petroleum Company of PNG (representing the government) and Mineral Resources Development Company (representing affected landowners).

Late in 2010 the PNG LNG Project was thoroughly underway. The Hides gas field was developed and readied for operation, a major airstrip was built at Komo south of Tari, and the massive pipeline from Hides and Angore to the coast and, thence, underwater to processing plants at Caution Bay, near Port Moresby, was completed. Once again, however, the Juha field was inactive. The pipeline that would link Juha to Hides was not in place. To local people—Febi and Kubo—this was frustrating and puzzling. Their expectations were not, as yet, fulfilled. It was, as so often it seemed to be, Papua New Guineans living elsewhere who benefited from development.

Exploration continued in the area and, from late December 2012 to May 2014, Suabi hosted base camps for the explorers. In the first few months of 2013, lines were cut through rugged limestone country north of Juha to facilitate a geo-survey (Fig. 3.2). From April to June of that year the same operators—Oil Search and Oilmin—initiated new seismic surveys in the Juha area. The survey lines cut through dense rainforest reached into Hela Province, and beyond the tree line. They ceased work in late June—bad weather was implicated in a helicopter accident—returned in September, moved their campsite to Juha and, with several hundred employees, continued the interrupted seismic project. The camp at Suabi was taken over, and enlarged, by Talisman who, for three months, undertook geo-surveys and seismic surveys on the western slopes of the Karius Range. Thereafter, to at least the end of April, the camp was put on hold—it was operated by Gama ProjEx[14]—while Talisman managers, based in Calgary, Canada, decided whether to continue work on the Karius Range, move elsewhere or find a new client for the Suabi camp. By late April the decision had been made and the camp was being dismantled.

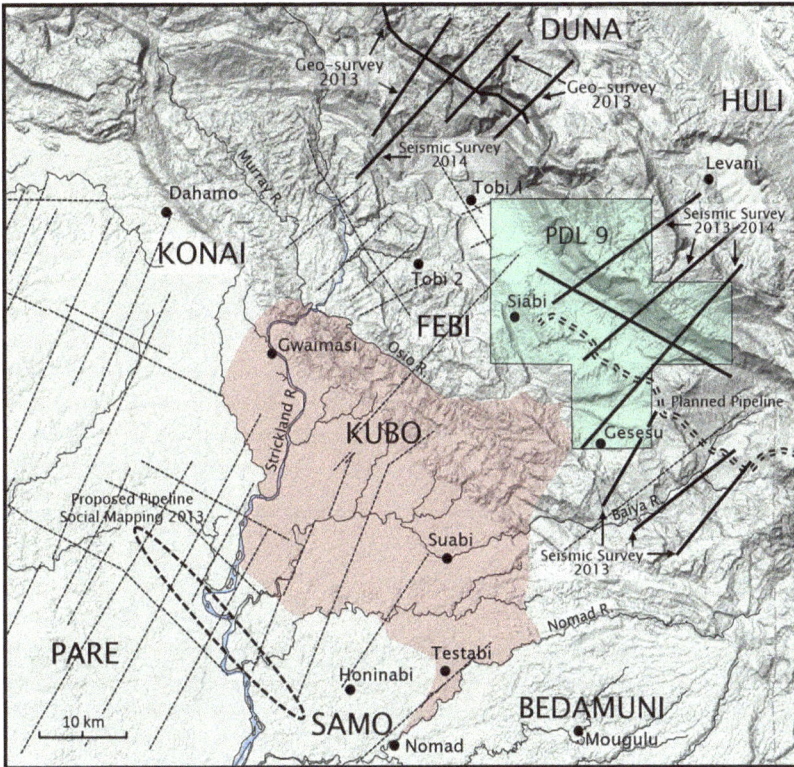

Figure 3.2: Map showing route of proposed pipeline and the location of exploration-related activities in the years 2012–14.

Sources: Based on Google Maps and authors' data.

Notes: Solid lines depict seismic and geo-surveys. The precise locations of earlier survey lines within and near Petroleum Development Licence area 9 (PDL 9) are not known to us. The fine dashed lines show the location of seismic surveys in the years preceding 1993. Some that cross Kubo territory extended further to the northeast than we have been able to show; of these, only the survey that crosses Osio River was conducted after 1986. The territory of Kubo people, as reported in 2014, and the area of PDL 9 are highlighted.

---

14   A description of a Gama ProjEx modular camp was viewed on 12 May 2015 at: technology. tki.org.nz/Resources/Case-studies/Technologists-practice-case-studies/Electronics-and-control-technologies/Modular-field-camp-system.

From June to mid-October 2013, a preliminary social mapping study and associated environmental studies were underway along the route of a proposed pipeline that would cut southeast from recently proven wells at P'nyang, in mountains to the northeast of Kiunga. The pipeline would cross the Strickland River just south of Kubo territory, near its junction with the Damami. Suabi provided a convenient location for refuelling and for assessing whether any Kubo people held rights to land in the impacted area. They missed out by about 5 km. And, again to the north, in the early months of 2014, from the Juha camp, a fifth seismic survey was initiated to the west near the Strickland.

From the perspectives of Kubo and Febi people, through these three decades, exploration activity was intermittent. The explorers sometimes found valuable resources and held them for the future. There was, it seemed, much secrecy. Eventually, however, those resources would be taken from the ground and royalties paid to land owners. In the meantime it was necessary to host the company representatives and, as possible, accept employment at base camp or in the field as labourers, assistant loadmasters, security officers, laundry workers, assistant cooks and so forth. Only one Kubo man had permanent employment associated with exploration activities, as a fully trained loadmaster with Pacific Helicopters. Kubo people, in particular, felt disenfranchised. They hosted the base camp. To their north, east and southwest, at the borders of their land, touching it but never on it, they saw and sometimes worked with exploratory ventures (see Fig. 3.2). Decades earlier a few lines had been cut through their own land but these had come to naught and the explorers had departed. Kubo people thought of themselves as being at the 'centre of Juha' but never received what they felt they deserved: mineral resources or gas on their own land and the unambiguous ownership rights that would follow.[15]

---

15   Additional activity, known to us, since mid-2014 is a 72 km seismic survey west of Juha close to the Strickland River and, in 2016, drilling of two wells associated with this survey and a well named Muruk 1 immediately northeast of PDL 9 (Oil Search 2015: 29, 2016a, 2016b, 2016c). The first of the Strickland River wells, close to the junction of the Liddle River with the Strickland, was plugged and abandoned after the Darai limestone was found to be water-bearing (Oil Search 2016a). The second Strickland well was plugged and abandoned in November 2016. In December 2016, Muruk 1 was reported to hold as much as USD14 billion of gas, providing the potential for an additional gas production train and enhancing the likelihood that the Juha wells will eventually be brought into production (Chambers 2016). In February 2016, Frontier Resources Ltd reported that it had been granted an exploration licence that covered areas at the headwaters of the Baiya

Geology was the problem. What the explorers wanted was in the mountains north of Kubo territory. The backswamps at lower altitudes offered little that sustained the interest of the mineral, oil and gas hunters. But some Kubo people saw it differently. Satan was implicated. He had stolen the valuable things—gold, diamonds, oil, gas—and hidden them deep in caves or in mountain hideaways where few people lived. White people had learned his trickery so this is where they searched. But there were deep caves that they did not know about. One day, properly equipped, with ropes for climbing and torches, local people would gain access and claim what was rightfully theirs.[16]

Petroleum and Energy

Patrol officers, missionaries, geologists, anthropologists, school teachers and community health workers may come and go—that is their fashion—but, when present, they are present in person. They and their impacts on local people are tangible. But the State—government bureaucracy—is distant and anonymous. It is the State, however, initially through Acts of Parliament, but subsequently through the practices of various national and provincial departments, that ultimately determines how, and to whom, royalties that flow from resource extraction ventures will be distributed. Multiple steps are entailed. They are complex, legalistic and seldom implemented in accordance with the letter of the law. And, further, the 'letter of the law' is itself subject to change.

The PNG *Oil and Gas Act 1998* requires that, prior to exploration, a preliminary social mapping and landowner identification study should be undertaken (Goldman and Ernst 2008: 10; Fitzpatrick 2010). This provides an initial frame for meetings of local-level stakeholders: a meeting at which all relevant stakeholders sign an Umbrella Benefits

---

and Nomad Rivers with a potential copper-gold-molybdenum target (viewed 30 April 2016 at: www.frontierresources.com.au/assets/ASX_-Muller_Range_17.2.16.pdf; PNG Mining Cadastre Portal, viewed 6 May 2016 at: portal.mra.gov.pg/Map).

16   Local people working on survey lines are advised by landholders to use a 'special language' when at higher altitudes in the mountains. In particular, some names must not be said; rather, to offset the likelihood of violent and terrifying storms, their referents must be alluded to by metaphorical equivalents (compare Franklin 1972). The belief that Satan had hidden valuable things, and that white people knew this, was reinforced in January 2015 when local men became concerned that the 'Devil' had infiltrated their mountain campsite and the white labour manager hired the Suabi pastor to perform what proved to be a satisfactory exorcism (Jim Savage, Willie Samobia, personal communication).

Sharing Agreement and a subsequent meeting at which landowners within each licence area are allocated their share of project benefits— the Licence-Based Benefits Sharing Agreement (LBBSA). Further, the PNG government prefers that benefits from extraction of oil and gas be distributed to legally recognised ILGs. To this end, it revised an earlier pre-Independence Act of the PNG House of Assembly and passed the *Land Group Incorporation (Amendment) Act 2009*, with a 42-page guide book and a 123-page training manual provided to facilitate implementation (GoPNG 2012a, 2012b). The revised Act came into effect on 1 March 2012.

In 2005 Sari Mora, accompanied by two community affairs officers from Oil Search, arrived at Suabi to conduct a preliminary social mapping study of people living in the Juha area. Prior to this year, oil and gas exploration had proceeded in the absence of social mapping. Mora's report provided a history of resource development activities in the area and discussed 'ethnic groups' (Goldman and Ernst 2008: 14). It is likely that at this time Febi people received advice with respect to setting up ILGs, for 11 applications had been lodged on 30 August 2005 (Ernst 2008: Appendix 3; Goldman and Ernst 2008: 170–3). Each of these carried the name of a purported clan: ten qualifying (by 2014) as Febi, one as Kubo.

In May 2009, the PNG DPE—'DP' to residents of Suabi—convened the Umbrella Benefits Sharing Agreement meeting for the PNG LNG Project at Kokopo, the capital of East New Britain. The meeting, which lasted for weeks, commenced with 150 hand-picked delegates but blew out to several thousand as people arrived from areas associated with the project (Fletcher and Webb 2012: 69; Garnaut 2015). Present were representatives of the National Government, four affected provincial governments, seven local level governments and project area 'landowners' from the multiple licence areas to be impacted by the PNG LNG Project. Their transport and accommodation costs were covered by the DPE.

The Kokopo agreement comprised 55 pages with an additional 69 pages for signatures (GoPNG 2009), though none of the ten Febi and Kubo representatives whose names we recognise have actually signed. Future benefits to landowners were to accrue from royalties, equity and a development levy. In addition, the State would set aside PGK1.2 billion for Infrastructure Development Grants and PGK120 million for Business Development Grants, with the latter accessible only by local landowners who had registered a company. Over the anticipated 30-year lifespan of the project, landowners were expected to receive more than 20 billion kina. These benefits were to be distributed differentially among landowners associated with licence areas and with well heads, buffer zones and pipeline easements. Details of how benefits would be distributed to individuals or groups within particular licence areas were not considered at the meeting.

In mid-November 2009 an LBBSA meeting was convened at Suabi. This was specifically concerned with potential beneficiaries associated with the area covered by Juha Petroleum Retention Licence 2. Key participants were the Independent State of PNG, two provincial governments, two local level governments and representatives of 18 named Febi and Kubo 'clans' together with six named 'clans' whose primary affiliations were with Huli and Duna-speaking peoples from what was to become Hela Province.[17] Some of these people asserted historical and mythological connections to the Juha area and argued that the Gesesu community, 15 km south of the Juha well heads, was mixed Febi and Huli (Ernst 2008: Appendix 4; Goldman and Ernst 2008: 98–9). The Hela participants were dissatisfied by the venue chosen for the meeting. It was too muddy, facilities and food were inadequate, women came to the meetings and there was risk of sorcery. Their arguments and insistence prevailed. On 3 December the meeting moved to more salubrious facilities at Moro, a long-established base for Oil Search, near Lake Kutubu in Southern Highlands Province.[18]

The Moro agreement stated that Juha area landowners would comprise Febi clans together with other Western and Hela Province clans 'as invited by the Febi clans'. In essence, those who signed the agreement were accepting that the PNG LNG Project was valuable, should proceed without impediment and that, to the date of signing, all that had happened was in accordance with the law. However, it was only with respect to CDOA equity that the agreement specified the distribution of future benefits to clans. This equity arises from a Coordinated Development and Operating Agreement that had been signed by PNG LNG Project companies in May 2008. The three clans considered to be most closely

---

17   Hela Province, formerly part of Southern Highlands Province, was formally created in May 2012. The Hides gas field, the most productive of the PNG LNG fields, and the gas processing plant at Komo are within this province. We use the name Hela for this region even when referring to years before 2012.

18   Our information about the Suabi phase of the LBBSA meeting and the Moro agreement comes courtesy of Anaïs Gérard, who was at Suabi for much of the meeting and who, on a later visit, photographed all pages of a copy of the agreement held by a Febi man.

associated with the well heads were to each receive 15 per cent of the CDOA quantum allocated to the licence area as a whole. The other nine Febi clans were to each receive 5 per cent, the six 'invited' Kubo clans would share 2 per cent and the six 'invited' Hela and Duna clans would share 8 per cent. Should Febi clans fail to agree to this proposed distribution before 15 January 2010, then the matter could be resolved by a gazetted ministerial determination. The Moro agreement also stated that DPE would coordinate a clan vetting task force that would 'identify, coordinate and facilitate incorporation of land groups for each affected clan or sub-clan identified within the licence area in accordance with the Lands Group Incorporation Act 1974'.[19]

The Moro meeting was an expensive exercise. Hela participants were flown first to Suabi and then, together with at least 59 local-level participants—including two church and three women representatives—were flown from Suabi to Moro. At the close of the meeting many of these people were returned to their home villages by either plane or helicopter while others were provided with transport and funds to attend meetings at Port Moresby. And the meeting failed in its primary purpose. Some people, who asserted they were landowners, fought with DPE officers during the meeting (Bashir 2010a), and the agreement itself, signed by local-level participants but not, apparently, by any government personnel, immediately generated controversy (Anon. 2009).[20] There was disagreement about the proposed distribution of benefits, there were assertions that the social mapping studies were flawed and incomplete (McIlriath et al. 2012), and there was much concern about the fact that the PGK11 million allocated to the Juha licence area for Business Development Grants had either not been released or had been corruptly mismanaged (Bashir 2010a, 2010b; Rouzet 2013).

---

19 'Clan vetting' was devised by officers of PNG government departments after the Kokopo meeting in an attempt to resolve perceived difficulties with social mapping and landowner identification studies that were required under the *Oil & Gas Act 1998* and were conducted as consultancies under contract to relevant petroleum companies. These studies did not name the individuals who might eventually be judged to be legitimate beneficiaries of royalties and other benefits and this proved frustrating to both bureaucrats and local people. The clan vetting process, conducted in the field by DPE officers, attempted to list all deserving clans and subclans together with their named representatives. The process has not, however, been implemented in accordance with government legislation and, in the years from 2010 to 2016, has been often challenged by the judiciary (Koim and Howes 2016).

20 The copy of the Moro agreement that we have seen was signed by representatives of all named Febi, Kubo, Huli and Duna groups but not by any government personnel.

At around the time of the Moro meeting a massive PNG LNG Environmental Impact Statement, including a 1,610-page social impact assessment, was finalised and submitted to the PNG Department of Environment and Conservation (Coffey Natural Systems 2009). In October 2009, this was accepted by the PNG government and, a year later, the PNG LNG Project was approved by participating partners. Soon afterwards, much of the Juha area was reclassified from Petroleum Retention Licence 2 (PRL 2) to Petroleum Development Licence 9 (PDL 9). The former classification applies to areas awaiting further appraisal, the latter to areas of intended resource extraction.

On 8 March 2010 Hon. W. Duma, Minister for Petroleum and Energy, signed an *Interim Determination of Juha Field (PDL9) PNG LNG Project Identified Landowner Beneficiaries*. Representations and lobbying by Huli and Duna speakers had been effective, for where, under the Moro agreement, they were to receive 8 per cent of benefits their allocated share was now 34.5 per cent. With few exceptions, the *Interim Determination* did not specify named groups within each of four broad categories of 'landowner beneficiaries'. That was yet to be achieved.

In December 2010, under the auspices of the Nomad Local Level Government (Western Provincial Administration), 36 Febi and Kubo men representing 12 clans met in Kiunga and, after a briefing by technical officers, continued to Port Moresby. Here they held meetings with both government and PNG LNG Project officers. Their aim was to counter representations by Hela people, finalise and sign a *Juha PDL 09 Landowners Declaration* and, once again, seek release of, or access to, funds that had been set aside for Business Development Grants. Under their declaration they agreed to 'honor the conditions of the Juha Petroleum Development Licence 09 License Based Benefit Sharing Agreement (LBBSA) that we signed in accordance with the Moro LBBSA Forum on the 08 of December, 2009 in compliance with the Oil and Gas Act'. There was no reference to the Minister's determination of March 2010. Their presentations, and those that emanated from Hela, reinforced the growing understanding that more work was needed before rightful claimants to the Juha area could be identified. But it was not until November 2013 that this process came to fruition. In that month DPE organised a clan vetting exercise at Siabi, the small Febi village closest to the Juha wells.[21]

---

21   In 2007 the population at Siabi was estimated to be 73 (Goldman and Ernst 2008: 32). By 2013, it was considerably less.

Most of the people who attended reached Siabi by walking, a minimum of two days through mountainous country from Suabi. As reported to us, DPE officers who ran the meeting sought to validate authentic landowners by eliciting knowledge of named features of the landscape, and paid minimal attention to assertions of mythological connection. They were aided in their endeavour by one employee whose paternal affiliations were with Febi.[22]

In April 2014, DPE released a *Juha PDL 9 Recognised Major Clan and Sub-clan List*. At the time of the Kokopo meeting it had been asserted that 11 Febi clans held rights over the Juha area. At the time of the Moro meeting, Febi representatives acknowledged 12 Febi clans, six Kubo clans and six others as legitimate beneficiaries. But much had changed. Where, previously, Febi disputed the legitimacy of Huli and Duna claimants they were now, additionally, engaged in redefining themselves. A combination of internal discord and strategic thinking (Minnegal et al. 2015) saw the number of clans judged to be eligible increase from the 24 listed at Moro to 53—of which, in our 2014 judgement, 23 are unambiguously Febi, eight are unambiguously Kubo, three are admixtures of Konai, Febi, Siali (Tsiani) and Huli and the remaining 19 are Bogaia, Duna or Huli (see Table 5.1). By late 2014, DPE had made no moves in the Juha area towards advising and assisting with the establishment of ILGs.

Entangled Histories

The histories of outside engagement with Kubo and Febi people by explorers, government officers, missionaries, and men seeking gold, oil or gas, overlap in time and in influence. They are not discrete. Each category of intruder draws on the experiences of those who preceded them. And, in turn, the people themselves seek ways to navigate the shifting terrain of the multiple ways of living in, and knowing, the world to which they are exposed. Through the 1960s and 1970s people living north of the Baiya River were seldom visited by agents of the colonial government. They were left alone. They themselves initiated changes that occurred,

---

22   The Febi man employed by DPE was listed as the representative of one subclan within his natal clan and as the representative of both a different major clan that was unambiguously Kubo and a subclan within that clan. The DPE employee's connections to the latter clan were through his mother and, in fact, his mother's brother would have had a greater claim to be nominated as the clan representative.

though sometimes at the prompting of government representatives. After Independence in 1975, there was even less contact. If people felt a need for government intervention or advice they were expected to report to Nomad; the officers based there seldom came to them. In the 1980s, it was missionaries and pastors, rather than government, who settled in these areas—ECPNG in western Kubo, ECPNG and SDA in eastern Kubo and the Christian Brethren Church at Tobi in northern Febi territory. It was missionaries who facilitated the arrival of education and health services, though in fact, north of the Baiya River, this occurred only at Suabi in eastern Kubo land. Though government did eventually begin to staff schools and medical aid posts, support for those staff was minimal, salaries often were not paid, and retention of staff unsurprisingly was difficult; missionaries continued to play a crucial role in the viability of those services.

From the mid-2000s, however, with the advent of what would become the PNG LNG Project, Kubo and Febi people were catapulted into worlds of bureaucracy, printed words and the law. They had no prior experience that might help make sense of these new worlds. Relatively few people could read. None could comprehend the language of Acts of Parliament or of documents generated by government departments. Indeed, in 2014, we met no one at Suabi who knew what 'percentage' meant. No one was sure what an ILG was or how one went about registering either an ILG or a company. But gradually, through that decade, it became necessary to traverse this terrain. The wells that had been drilled at Juha held the promise of great wealth for, at least, some people. To ensure access to that wealth—to ensure legitimacy—it was necessary to comply with all that it seemed government demanded. And it was necessary, too, to discover or invent connections of kinship and history with those who, it seemed, had unambiguous rights to that wealth. For ten years people were confronted by a 'promise' that seemed never to be fulfilled. For ten years they navigated a barely comprehensible, and ever-shifting, terrain that led to the future. What they did is the subject of later chapters.

4. Suabi: 2011–14

Jeff returned from work soon after 7 am. Through the night he had been a security guard at the exploration base camp. Now he was tired and slept. In mid-afternoon, he joined a group of young men and boys in a small, high-set house where they played cards and gambled for money. He had seen little of his wife Alice, or their eight-month old daughter Kamari, through the day. Alice was exhausted. As happened every day now, she had washed nappies and clothing in the river, hung them out to dry and cared for the child. She had little relief from these chores. She was six or seven months pregnant, emotionally overwrought and feeling abandoned. With Jeff and Kamari, she was living with his parents but did not get on well with them. Nor, with Jeff working, could she contribute as much to daily subsistence tasks as was expected.

In the late afternoon Alice, carrying Kamari, started to fence a small household vegetable patch that she and Jeff had planted several months earlier. But the fence was unnecessary. Alice was making a public statement, showing those nearby that it was she who supported the family and that she did so without assistance from her husband. She worked, while Jeff slept and played cards. He did not harvest food from gardens, or hunt, or fish to support his family.

Alice and Jeff had been having difficulties for some time. Jeff had been angry when their child was born. He had wanted a son, blamed Alice when she gave birth to a daughter and talked of finding a second wife. Today their growing difficulties spilled over. Jeff saw what Alice was doing, raced down from the house where he was playing cards, pulled Kamari away, thrust her into the arms of his youngest sister, ran back and threw Alice to the ground. She fought back and keened: high pitched, almost unintelligible as she gave vent to her complaints through tears. He didn't help with the work, he gambled, he didn't share his money with her. She grabbed a stick, tried to hit him, picked up cooking stones from the pile outside the house and threw them at him. He danced out of the way, laughing at her. He ran at her again, pushed her and became furious. He punched and kicked the trunk of a banana plant. People gathered, watching, increasingly concerned as the fight escalated. Two older women moved in, standing between them, holding them apart, wresting sticks and stones from Alice's hands. Paul appeared—he was the locally-appointed Law and Order Committee—and ordered them to stop. Still keening loudly, Alice retreated to her house. Other people moved to a different house. They discussed the problem. 'Cards,' they concluded. Jeff was playing cards and Alice was 'jealous'. Gisio found the pack of cards and threw them into the fire. 'Cards always cause trouble,' she said. Now they have caused a major problem in this part of the village. She told her own small children to leave: 'There is nothing to see here,' she said.

Alice wept through much of the night. The next day, Jeff returned to work. One foot was bandaged. He was limping and using a stick for support. His mother was sitting with us as he limped past. Her child's wife caused the injury, she told us. '*Sobo bami*' [bad woman], she declared vehemently. But Jeff was not hurt; his injury was fabricated. Within a week they were fighting again.

It was in March 2014 that Jeff and Alice fought. Twenty years earlier tensions between husband and wife rarely escalated to physical fighting. At Gwaimasi and Mome Hafi, during five visits amounting to 25 months, there was only one occasion when a man struck his wife. He had been hunting for several days, failed to kill a pig, felt that he was ageing and was no longer capable, returned from the forest, sat moodily at his hearth and, irritated with himself, deliberately broke an arrow to mark that his hunting days were over. His wife comforted him. It was just bad luck, she said; he would kill a pig next time he went hunting. He did not want comfort. He hit her, a single blow and was deeply shamed. Now, however, people encountered different pressures. Population growth, aggregation into larger and more sedentary communities, emerging constraints on access to land, changes to marriage practices and expectations, paid work and its effects on both access to money and the distribution of subsistence tasks among men and women, were all implicated in the deterioration of the relationship between Alice and Jeff. These are themes that we shall take up in this chapter where we depict life at Suabi in the years 2011 to 2014 and reflect on changes through the preceding two decades.

Place and People

The village now known as Suabi came into being in the early 1980s. The missionaries Tom Hoey and John Fletcher chose a site on a bend of the Baiya River as the location for a new airstrip, and encouraged people from nearby longhouse communities to help build it. Through the next few years, as the airstrip became a reality and John Fletcher and the Bedamuni pastor Fiagone settled nearby, increasing numbers of people relocated from previously scattered longhouse communities so that, by the mid-1980s, 150 people identified as residents of the new village. Most were Kubo-speakers, some were Febi-speakers and perhaps 50 were Bedamuni-speakers who, attracted by the prospects of the airstrip, had relocated from the southeast and established a cluster of houses on a ridge overlooking the river. At first, most of the Febi-speakers were men who in earlier years, sometimes as children, had moved south from the mountains, married Kubo women and, for the most part, lived as Kubo. Other Febi-speakers came later—and, indeed, continue to come—to access the school, health services and communication facilitated by the presence of the mission station. Bedamuni-speakers returned to their own land in the late 1980s, after a dispute, but re-established a community a few kilometres from the airstrip about 20 years later.

Table 4.1: Population census of Suabi (March–April 2014).

CORNERS	S	O	T	S & G	G & D	B	TOTAL
Households	18	21	25	9	14	10	97
Married ♂	21	23	28	14	19	10	115
Married ♀	21	26	31	14	21	10	123
Widower	1	–	1	1	1	–	4
Widow	3	8	3	1	4	3	22
Divorced ♀	2	–	–	2	1	–	5
Unmarried ♂	62	64	65	29	51	23	294
Unmarried ♀	33	45	43	26	41	13	201
TOTAL	143	166	171	87	138	59	764

Source: Authors' work.

Notes: Separate records are provided for six discrete assemblies, comprising eight Corners, within the greater Suabi community. Corners are coded as S = Suabi; O = Owabi; T = Timaguibi; S & G = Sasosi & Goonabi; G & D = Gweobi & Dufosi; B = Bigusubi.

As the number of people living at Suabi grew, and the village increasingly assumed an air of permanence, earlier structuring principles of exchange marriage and co-initiation that had patterned co-residence weakened. Within the greater Suabi area, people established household clusters that, in large part, were formed on the bases of language and both clan and mission affiliation. (In this chapter, and later, we follow contemporary practice at Suabi by using the word 'clan' rather than *oobi* to refer to group affiliation, a shift that reflects both the penetration of government discourse, the increasing use of Tok Pisin in the community, and the multi-lingual composition of the community.) By April 2014, there were 764 people residing at Suabi, together with one community school teacher with his wife and son, and seven boys and three girls from other communities who lived with families at Suabi while attending elementary or community school. The ratio of males to females was 1.18.[1] These people were dispersed as six residential clusters, locally termed Corners, though some people talked in terms of an even finer classification (Table 4.1, Figs 4.1, 4.2).[2]

---

1   This value will slightly understate the bias in that more males than females have permanently departed from the community. Records of 103 births at the Suabi Community Health Centre for the years 1999 to 2015 yielded a male-biased sex ratio of 1.26. Our own records from 1986 to 1995, for a small sample of 23 cases where date of birth is known and underreporting of females can be excluded, yielded a male-biased sex ratio at birth of 1.3.

2   We provided five people with forms on which, with the assistance of others, and for each household in one or two Corners with which they were familiar, they recorded the names of resident married men, widowers, married women, widows, divorced women, unmarried men/boys, unmarried women/girls, child-boys and child-girls. 'Residents' were classed as those whose primary residence was said to be Suabi. Our assistants decided who did and who did not qualify as a resident—they had very firm opinions—and recorded the location of people who were absent at the time of the census. We cross-checked completed forms against previously recorded family lists, school enrolment lists, community health centre records and long-term diary and kinship records and asked our assistants, and others, for clarification wherever there was ambiguity. This was particularly important because people have, and use, multiple names. It was also important to eliminate multiple listings of the same person. Some young males and females regularly move between Corners of the greater Suabi community and some were listed more than once or forgotten. With these very mobile individuals we decided which Corner qualified as their primary residential focus. Despite our guidance regarding the distinction we sought between unmarried man/boy and child-boy, and between unmarried woman/girl and child-girl, there was little consistency in the judgements made by our assistants. In Table 4.1 we have pooled these categories as, respectively, 'unmarried male' and 'unmarried female'. The PNG National Census of 2011 recorded population data for Corners of Suabi south of the Baiya River—pooled as Timaguibi—but not for those north of the river. Census officers visited the market at Suabi Corner, explained what they wanted and were told forcefully that 'government' always came wanting something from the people but never gave anything in return. People north of the river refused to cooperate and, thus, no population statistics were collected from Suabi, Owabi, Gweobi, Dufosi or Bigusubi Corners.

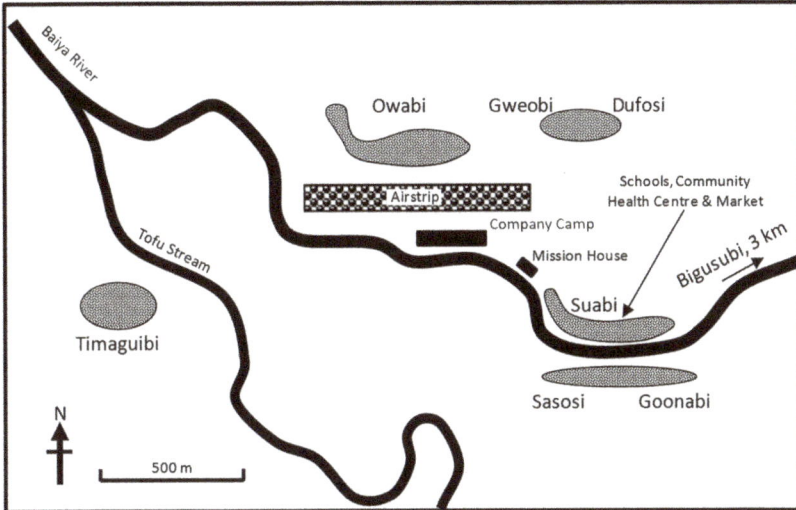

Figure 4.1: A schematic representation of the greater Suabi community in 2014.

Source: Authors.

Note: Clusters of houses–'Corners'–are shaded. Direction and distance are approximate.

Figure 4.2: Entrance to Owabi Corner.

Source: Photograph by Peter D. Dwyer, 2011.

In broad outline, though with the exception of Suabi Corner, the affiliations of nearly all senior males at each Corner reflect an orientation of residence to the location of the land of their clans: at Owabi to the west and northwest, at Timaguibi to the west but south of the Baiya River, at Sasosi to the south and east, at Gweobi to the north and at Bigusubi to the east. Predominant language affiliations are Kubo for Owabi, Timaguibi, Sasosi and Goonabi Corners, Febi for Gweobi and Dufosi Corners and Bedamuni for Bigusubi Corner. Suabi Corner is more diverse than others with respect to language affiliation. In 11 households the senior resident was Kubo, in four he was Febi by birthright though with long-standing Kubo affiliations, and in three others he was an outsider from, respectively, Bedamuni, Oksapmin and Morehead. Primary mission affiliation also maps onto residence. Residents at Suabi, Owabi, Sasosi, Goonabi and Bigusubi are, for the most part, aligned with the Evangelical Church of Papua New Guinea (ECPNG) though at Owabi, since the mid-1980s, one cluster of households is aligned with the Seventh-day Adventist Mission (SDA). Timaguibi was established as an SDA community in the mid-1980s, though that connection has been diluted through the past decade, and the predominantly Febi Corners, Gweobi and Dufosi, are aligned with the Christian Brethren Church. By 2014, though no one expressed disbelief or doubt in relation to the Word of God as represented by the Bible, there were increasing numbers of people, especially of men in their later 20s and 30s, who rarely, if ever, attended church services.

The size of the aggregate Suabi population in 2014 is, in part, the product of more people settling there on a permanent basis and, even more so, the product of rapid population growth arising from increased infant survival and increased longevity. Our tentative estimates of annual population growth are 3.6 per cent from 1987 to 1999 and 5 per cent from 1999 to 2011.[3] The latter figure approaches the high of 5.4 per cent recorded for the Southern Highlands Province in the 2001 National Census. Since 1986, the Kubo population has doubled to approximately 1,000 people. Between 1999 and 2011, the number of living children per female increased by

---

3   Our estimates of population growth rate draw on data from people who were resident at Gwaimasi in January 1987 and separate the periods 1987–99 (population size 21 to 32) and 1999–2011 (population size 32 to 57.5, adjusting for cases where a man or woman moved into or out of Gwaimasi or, after Gwaimasi was disbanded, Mome Hafi). Data on the numbers of living children or grandchildren per female are based on records from 1999 and 2011. Sample sizes for females include only living, married females for whom details are complete. For living children per female, n = 27 and 38 women respectively for 1999 and 2011; for living grandchildren per female, n = 27 and 35 women respectively for 1999 and 2011. One older female who lost her first and only child and one nulliparous female are included in the samples.

a factor of 2.37, with some women now having eight or nine living children, and the number of living grandchildren per female increased by a factor of 3.0 with some women now having seven or eight grandchildren.[4]

Multiple factors are implicated in the increases in both family sizes and longevity. These include abandoning customary birth-spacing practices (progressively since 1987), abandoning infanticide of newborn babies born to unwed women (since 1987), abandoning the killing of persons identified as sorcerers (largely in place by 1987), and, since the mid-1980s, access to at least low-level medical care that has included missionary-sponsored emergency evacuation, the presence of a trained community health worker at Suabi, availability of community health centre pre-natal and post-natal care leading to increased survival of both mothers and infants, intermittent inoculation for infants, and treatment of malaria and a variety of infections. In recent years, mission doctors based at Rumginae have, from time to time, visited the village to conduct clinics. In 2004 a team from Australian Doctors International conducted clinics and distributed mosquito nets throughout the Nomad region, though many of the nets have been used for fishing and not for their intended purpose (Bowman 2004; Gentleman 2015; Minakawa et al. 2008. See also GoPNG 2005; JTA International 2009). And, in late April 2014, in response to an outbreak of measles in the Upper Fly River area, two paramedics visited Suabi to vaccinate people between the ages of six months and 20 years. They combined this program with another in which they vaccinated all females of reproductive age against tuberculosis and presented talks in which they spoke of hygiene and sexually transmitted diseases but not birth control.

But not all those at Suabi through recent years were from local language groups. One man from Oksapmin and another from Morehead married local women while based at Suabi as community health workers; one returned to raise his family there, and the other visited often to spend time with his wife's family. Three women, one from Wabag, another from

---

4   Body size has also increased through the past two decades. Many young men and women in their 20s are taller than their same-gender parent. Increases in height, weight and Body Mass Index have been reported from other PNG populations but, though presumably linked to socio-economic change, key causal factors have not been unambiguously identified (Ulijasek 1993, 2003; Norgan 1995; Adhikai et al. 2011). Some local people assert that these changes had been prophesised by members of Nomo clan: a new kind of people would come, people who were 'naked' and did not have their noses pierced and, when they came, local boys and girls would grow faster and larger; they would stop respecting their parents, the girls would grow breasts when very young, start having sex too soon and marry before they were ready.

Ialibu, and a third from Kiunga had married men from Suabi, and returned with them to live at Suabi. Through 2013 and into 2014, when intensive exploration was underway in the nearby mountains, a handful of white men—camp managers, geologists, helicopter pilots—and a hundred or more nationals flown in from distant parts of Papua New Guinea (PNG), were living in the base camp near the airstrip. This activity attracted others, who made their own way to Suabi in the hope—usually unmet—of finding employment as labourers; a few young men from Ialibu, Mount Hagen and Goroka spent several weeks at Suabi through early 2014, invited by local youths who knew them from school. Though occasionally judged to cause trouble, they were nevertheless welcomed, seen as reflecting a long-awaited recognition by the wider world of Suabi's importance.

Livelihood

Gardening, sago processing and hunting and fishing remain primary as life-support systems at Suabi, though changes have occurred in the expression of all these practices. Bananas continue to be the primary carbohydrate food produced at gardens but, relative to earlier years, sweet potato has become more important and the diversity of greens has decreased with *abika* (*Hibiscus manihot*), which is eaten in quantity on many days, the most important of these. Bananas and tubers require different modes of production and, for the most part, are grown at separate gardens. The yield from banana gardens is poor from about 20 months after planting (Dwyer and Minnegal 1993). At Gwaimasi, in the 1980s, banana gardens were abandoned after this time and sites left for 15 to 20 years before the well advanced regrowth was felled and the area replanted. At Suabi, people continue to take what they can for a longer period.[5] But the outcome of increased population and sedentisation at Suabi has led to a 12-fold increase since 1974 in the area of gardens and secondary growth that encircles the village site (Fig. 4.3). To this time, most people have been establishing new gardens slightly further from the village than their previous gardens. Most of these are within a zone that is currently recognised as being open to access by all. There is only one area, towards Wa River, where Suabi residents have consistently been gardening more than a few kilometres from the village.

---

5   Shingo Odani (2002), working with Bosavi people of the Great Papuan Plateau, reported that banana production measured as kg/ha/year was reduced by as much as two-thirds when gardens were maintained beyond the time that the first harvest was completed. (See also Kuchikura 1995.)

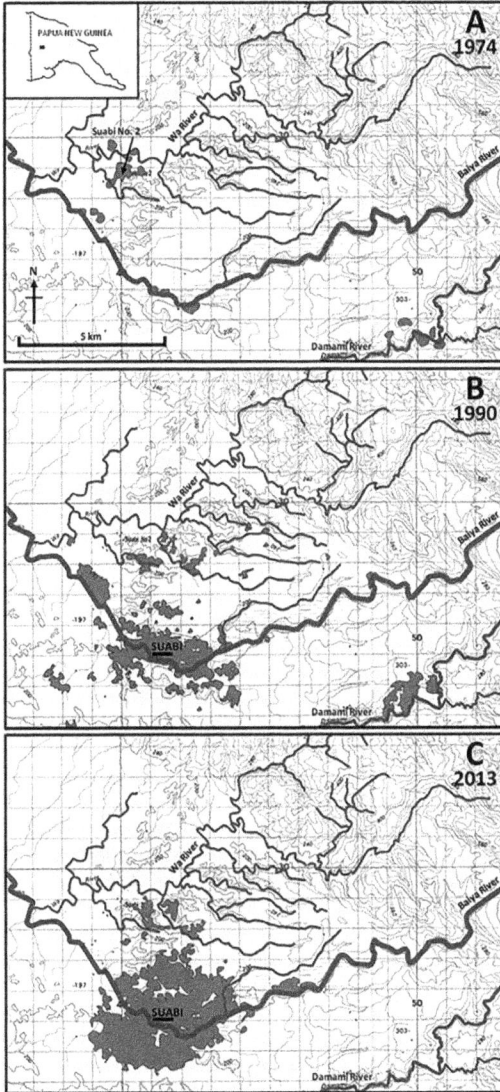

Figure 4.3: Extent of gardens and secondary growth in the area near Suabi, 1974 to 2013.

Source: The base map is taken from the PNG 1:100,000 Topographic Survey, Sheet 7385 (edition 1) Series T601, printed 1979.

Notes: Shaded areas show gardens, secondary growth and, in B and C, clearing associated with Suabi village and airstrip (solid bar). A: Shaded areas based on aerial photography at 1974. B: Shaded areas based on Landsat image at 30 December 1990. C: Shaded areas based on Google Earth image at 10 April 2013. Some areas of secondary growth shown on B do not appear on C; in part, this is because secondary growth becomes less easy to distinguish from primary forest after about 20 to 25 years; in part, it reflects differences in the quality of the Landsat and Google Earth images.

As yet, few earlier sites are suitable for replanting. Indeed, harvesting for longer than 20 months has the outcome that replenishment of soil nutrients takes longer and, hence, a longer fallow is necessary if yields are to be rewarding. It is likely to be for this reason that sweet potato has increased in importance. Sweet potato gardens must be fenced to minimise the depredations of wild pigs but this increase in labour is compensated for by the fact that a fallow period of only five to ten years is needed and that, with some composting achieved by burning before planting, production can be relatively rapid and high. At all Corners small household gardens are common but, unless composted, yields are poor. Old house sites, burned after salvaging usable building materials, are used as highly valued household gardens; they are very productive, though for only a single planting. Most vegetable waste from households is thrown into clumps of banana plants that then yield well. Coconut palms, which were not grown by Kubo before colonisation, are now abundant in the village area—some, in fact, are old and no longer yield—as too are okari nut trees (*Terminalia*).

When people began to settle at Suabi they were, initially, reliant on sago that they processed from palms that grew in swamps nearby. These were primarily wild, rather than planted, palms and, had people been reliant on them alone, they would not have met the requirements of the growing population. In the first decade or so after settlement people often travelled some distance to find suitable palms, returning to places they had recently abandoned to utilise palms that they themselves had planted. From the outset, however, they planted palms at Suabi, close to the numerous streams that drain to the Baiya River. It was 15 to 20 years before these palms would flower; the ideal time to fell a palm and extract the flour is shortly before it flowers and the starch stored in the trunk is redirected to flowers and seeds (Dwyer and Minnegal 1995). By the 2000s, then, there were many mature and maturing palms available close to where people lived. In the years 2011 to 2014 it was only when feasts were pending, and large quantities of sago flour were needed, that it was necessary to travel some distance, and sleep away from the village, to access suitable palms.

Wild animal foods remained important at Suabi, though people often complained that they were hard to come by and said that those living near the Strickland River ate much better than they themselves did.[6] Hunting was less frequent than it had been at Gwaimasi and men who did go hunting—it was usually older men—were seldom away for more than a day or two. Wild pigs and cassowaries were now more often taken by snares than by hunting with dogs or ambushing from hides, and men who had set a snare would leave the village at dawn and be back within an hour or two. Snares had been introduced in the mid-2000s by outsiders employed by companies exploring for oil and gas. They proved very popular and were often successful. Many fish, crayfish and prawns were obtained by netting, trapping, spearing and line-fishing in the river and streams and, as in earlier years, beetle larvae were incubated in the trunks of felled sago palms. Smaller items like fish, crayfish, prawns, frogs, lizards, snakes and birds—the last shot by arrow or, more commonly now, catapult—were often eaten privately, either near the place where they were taken or away from sight, after dark, inside houses. Indeed, at Suabi, most houses—including separate kitchen houses—were enclosed, or built such that open cooking areas faced away from walking paths. They were structurally not welcoming. In the circumstance of a large resident population sharing was less common, and captured animals were more often concealed from view. For these reasons, people who may themselves have eaten meat quite often might say that animal foods were scarce.

Pigs, dogs, chickens and cassowaries are kept as domestic animals by people at Suabi. In contrast to practices seen at Gwaimasi, pigs are now managed in such a way that they are tame to everyone; though older animals may be released into the forest to forage or, if sows, to mate with wild boars, and some larger animals are penned, many roam freely though the village area. They are fed on cooked bananas in the early morning and in the evening. They are killed and eaten at feasts, to mark weddings, as compensation payments, at the conclusion of many locally held court cases (see Chapter 6) and sometimes because people from one Corner—usually young men—are 'hungry' for pig and raise enough money to buy

6   During four-and-a-half months in 2013–14, the 143 residents at Suabi Corner publicly butchered, cooked and distributed ten wild pigs and 12 domestic pigs. It is likely that some additional wild and domestic piglets were killed and eaten privately. These values translate as 0.015 wild pigs and 0.02 domestic pigs per person per month. Comparable values from Gwaimasi in 1986–87 are much higher, especially for wild pigs; respectively 0.22 and 0.03 (25 people, 15 months, piglets excluded; Dwyer 1993: 131, Dwyer and Minnegal 1991: 192).

an animal from an owner living at another Corner. Through the years 2011 to 2014 the numbers of village-based pigs varied greatly, because late in 2011 people were attempting to maximise their holdings to satisfy the perceived requirements for a spectacular three-Corner feast a year later; a three-day feast at which 65 domestic pigs, together with dried meat from hunted animals, were distributed throughout Suabi and to several hundred guests from elsewhere. Dogs were by no means as abundant, relative to the number of people, as they had been at Gwaimasi in earlier years. When men did go hunting they called up four or five dogs to accompany them. These, however, were not necessarily their own dogs. Because hunting with dogs had decreased in frequency, and because men could call on other people's dogs, there was less reason for each family to keep many animals. There were about ten to 20 dogs associated with each Corner.

A few families reared cassowaries that had been captured in the forest as chicks. The animals were caged and considerable effort was entailed in provisioning them, on a daily basis, with diced pawpaw or, in season, tree fruits such as Java Apple (*Syzygium*). They were important in facilitating exchanges both between members of the SDA community—who neither ate nor raised pigs—and between people affiliated with other missions and SDA members. Finally, there were many chickens that ranged freely at all Corners. They were individually owned—sometimes by infants—and named, and often marked with a piece of coloured flagging tape tied to a wing. Eggs were seldom eaten, and mortality of chicks was high as a result, particularly, of hawk attacks and, less often, predation by pythons, dogs and pigs. Surviving young chicks were individually tamed and, from time to time, fed termites. Adult chickens, more often roosters than hens, were killed and eaten, sometimes after being purchased, to celebrate minor events such as the expected departure of a family member or, in one case, a child's birthday. Killing always entailed a hilarious chase through undergrowth and numerous attempts to shoot the bird with an arrow.

While the vast majority of food consumed at Suabi continued to be locally produced, the rice and tinned fish that, through the 1990s, had become expected at intercommunity feasts by people at Gwaimasi have become a much more regular part of the diet. A shared meal to mark completion of a communal task—roofing a new house, or clearing the remains of a fallen tool-shed at the airstrip—would often now include these 'modern' foods. And those with money would occasionally purchase rice, tinned fish and noodles for an evening meal, eaten behind the closed doors of a family house. In terms of quantity, however, these foods contributed little to the overall diet. When available and affordable, people ate what they managed to procure immediately. When, as happened once a month or so, the trade store owner flew in a supply of rice the 200 kg that arrived was often sold, and consumed, within a day or two. Only those few who secured work at the base camp or on exploration lines through 2013 ate rice every day. But that, of course, simply added to the status of this food as epitomising modern lifeways.

When exploration companies were based at Suabi through the late 2000s, nearly all adult men and some women were employed and money flowed freely. 'We ate rice every day' was the primary memory people recited of that time. Local people began to experiment with growing rice themselves, and one man purchased a rice mill that, for a small fee, people could use to husk their harvest. But by 2011 the mill had broken; it had not been maintained. People now had to carry their crop to Nomad, a long walk with a heavy load that might result in a few cups of rice at most; the effort was 'not worth it', we were told. With no way to mill rice, people left seed-stock to rot. But the possibility of again growing rice remained of great interest, and possible ways to procure another rice mill were often discussed.

Involvement in wage labour for exploration companies had other effects, however. Through much of 2013, most men were again employed as labourers on a seismic survey near the headwaters of the Baiya River. Without men available to assist with building fences, no new tuber gardens were established through that time. By late December 2013, almost no sweet potato was available for harvest. Once the survey was completed, however, most families turned to clearing and fencing new gardens. The result was a synchrony in gardening and sago processing across the community, in marked contrast to the flexibility seen at Gwaimasi. With exploration work itself seasonal, constrained by the clouds that descend on mountains to the north and east from May through to September, a de facto seasonality of labour that had little to do with the local climate was now shaping patterns of food production and consumption at Suabi.

Movements and the Outside World

The pattern of movements at Suabi was vastly different from that observed at Gwaimasi two decades earlier. At Gwaimasi in 1986–87, people spent 21.9 per cent of available nights at small garden, sago or hunting houses within the local subsistence zone, 6.6 per cent of nights at or near communities within a day's walk of Gwaimasi and 11.5 per cent of nights further afield with nearly all these at Suabi or Dahamo. (These values exclude children.) At Suabi in 2013–14, again excluding children, people spent only 6.9 per cent of available nights sleeping at bush houses within Kubo territory. They spent 4.7 per cent of available nights sleeping at, or on the land of, neighbouring Kubo, Samo or Febi communities— some entailing an overnight stop en route—and 19.2 per cent of available nights further afield at Kiunga and nearby towns or at Highland and coastal towns, including Port Moresby.[7]

The huge reduction in absences at bush houses was unexpected. Given the great extension of secondary growth reaching out from Suabi (Fig. 4.3) we had expected that people would spend much time away from the village to satisfy subsistence needs. But this was not the case. Indeed, the reduction is greater than the values suggest because nearly half the recorded absences were associated with employment when a group of people, supported by an outside grant, contributed to upgrading part of the walking track between Suabi and a Febi community to the northeast. When a family made a sweet potato garden that had to be fenced then they might spend a week or more away from the village and, from time to time, a family or group of families would camp in the forest to fish and hunt because this was enjoyable; the mood was that the time away qualified as 'holiday' rather than necessity. For different families, however, these activities were seldom synchronised, with the outcome that our 137-day sample reflects an uneven distribution of subsistence-related absences from Suabi Corner. At Suabi the distances travelled to gardening, sago processing and hunting sites were often comparable to those observed at Gwaimasi, but while residents of the latter village would remain away for the period entailed by the task, those at Suabi often moved between village and work site on

---

7   Absences at other Kubo communities and at places beyond Kubo territory were monitored for 17 of the 18 households at Suabi Corner (74 people, 10,138 available nights). We were unable to satisfactorily monitor absences from the Bedamuni household. Absences at bush houses within Kubo territory were monitored for seven households at Suabi Corner (35 people, 3,857 available nights).

a daily basis. They would return on dark, or after dark, not infrequently in the rain. They preferred to sleep at home. They did not want to miss the central possibilities provided by Suabi—news or visitors from the outside.

Absences from Suabi in 2013–14 that were to, or near, the land of neighbouring Kubo, Samo or Febi communities were, proportionately, somewhat less than analogous absences from Gwaimasi. Moreover, while the vast majority of the latter concerned customary social interactions, less than 10 per cent of Suabi absences were motivated in the same way. Rather, employment at Juha and mission-sponsored functions at Honinabi accounted for most of these absences. Much of this difference between Gwaimasi and Suabi may be attributed to the large size of the latter community, in that many social obligations were fulfilled in situ— as, for example, exchanges and marriage negotiations between residents of different Corners. In addition, however, there was an expectation on the part of the Suabi community that people would come to them; an expectation that was, in a sense, reciprocated by the attractions Suabi held for those living at a distance.

In 2013–14, people living at Suabi were travelling further afield than was the case in the earlier period at Gwaimasi, and were doing so more often. Access to money, the value placed on formal education, relocation of kin and the desire, or felt necessity, to pursue 'roads to money' go some way to explaining the difference. About half these absences were to Highland and coastal towns (Mount Hagen, Goroka, Lae and Port Moresby) and half to Kiunga and nearby towns (Tabubil and Rumginae); at the time of our census, between 45 and 50 people classed as residents were absent at one or other of these locations. All absences in the first set of destinations were by males, with the vast majority by unmarried older youths and young men who were, purportedly, furthering their education.[8]

Absences at or near Kiunga were for diverse reasons, including education (70 per cent of records), health, school and mission administration, 'private business' and visiting kin. One young woman attended high school at Kiunga, another visited for reasons of health, one married woman visited

---

8   It was widely known—and, indeed, some of the youths themselves said as much to us—that some of these purported students, who had previously been at school, no longer attended; they used stated intentions to do so as an excuse to return to these distant towns because they had developed a preference for life beyond what they saw as the confines of Suabi. Some of these young men told us that they no longer liked the bush, that they did not 'know' it and did not 'know' how to hunt. A few stated strongly that they had no intention of returning to live on a permanent basis at Suabi.

to take a child to hospital and another accompanied her husband during a period of employment. These four records account for 26 per cent of absences at or near Kiunga. Men sometimes gave 'private business' as the reason for their absence (10 per cent of records). Sometimes they were purchasing stock for local trade stores, but most often they were intending to negotiate with government officials or with 'knowledgeable' Kubo or Febi men in the hope of receiving advice about accessing a Business Development Grant or registering an Incorporated Land Group. To people at Suabi, Kiunga was a busy and attractive place—for some men the temptations of women and alcohol were always in mind—and many visits were, in part, to visit kin and friends.

Visits to Kiunga were made easier for people from Suabi because, for a decade or two, there has been a well established 'Suabi Corner' in that town. In 2014, this comprised nine houses. The primary occupants of four of these are Kubo families, from Suabi, that relocated to Kiunga as much as ten years earlier. The primary occupants of two others are Febi families in which the senior male moves back and forth between Kiunga and Port Moresby and, only rarely, visits communities on either Kubo or Febi land. Between 25 and 30 Kubo people, together with about ten Febi people, are established as Kiunga residents. The remaining houses, commonly referred to as 'Boys' Houses', are available to other Suabi people when they visit Kiunga.[9]

Increasingly then, for people at Suabi the movements that mattered were those that reached beyond the world of local subsistence. For Gwaimasi residents in earlier years, the bush was a place to escape the tensions of village life, and journeys beyond immediate neighbours had been primarily to render life at the village sustainable. For many of those at Suabi, however, it seemed the bush was now somewhere to go only if necessary, and usually for as little time as possible. Attention remained focused on the village, and the opportunities available there to remain in touch with—and perhaps journey out into—the wider world. It was to Kiunga or the Highlands, and not to the forest, that many people now looked for

---

9   Our discussion of movements has relied on quantitative data from one Corner at Suabi. With three exceptions we think the sample is probably representative. At Bigusubi—the Bedamuni Corner—it is likely that there were numerous reciprocal visits to and from Bedamuni territory that we were not aware of. At Gweobi and Dufosi—the Febi Corners—there were more 'business-related' absences at Kiunga and Port Moresby than at any of the other corners. And, finally, at Owabi, Timaguibi and Sasosi there were one or two families who visited their own *oobi* lands more often, and for longer periods, than did any families from Suabi Corner.

escape from the tensions—and, indeed, boredom—of village life. Yet, for nearly all, travel out into that world, and the networks established with people far beyond the lands of Kubo and their immediate neighbours, continued to be framed in terms of finding ways to draw the resources and excitement of that distant world back to the local community—securing the means to make Suabi itself 'modern'.

Emerging Institutions

Kubo people valorise autonomy, the ability to 'decide for themselves' what they will do, and where and when they will do it. As a community, too, Suabi is now largely autonomous, though not financially independent. People here identify as citizens of PNG, and vote in local, provincial and national elections. The national anthem is sung at school graduations, and Independence Day is celebrated each year. But the various levels of government have little salience in day-to-day life. The nearest police are at Kiunga, and rarely respond to appeals for assistance in dealing with crimes. Politicians are known by reputation, but the only time in recent years that any have visited Suabi was in 2009, when the Licence-Based Benefits Sharing Agreement was being negotiated for the Juha gas field; though Suabi residents considered this a validation of their importance to the nation, there was no attention to local issues during that visit. Health and Education departments provide minimal oversight of locally-based staff, and there is little sense that people at Suabi are accountable to these external authorities; like 'government' itself, they are seen as potential sources of funds rather than as authorities to which people must answer.

Despite this, however, the notion of 'government' plays a crucial role in local imaginings of what it is to be modern. Its forms and its functions are replicated locally. There is an elected councillor, committee representatives for the different Corners, and two 'Law and Order Committee' men responsible for dealing with transgressions north and south of the river respectively. There is a head teacher in charge of the elementary schools who oversees three other teachers, a school board, a community health worker and associated Hospital Committee, a one-man Five-year Planning Committee and a one-man Market Committee. There are pastors, church management committees and Fellowship groups for women and youth, each with their designated leaders. Mission Aviation Fellowship had a designated local agent. And, in 2012, one man was standing for the position of President in the Nomad Local Level Government.

But though they had been accorded the 'title' and, not infrequently were both referred to and addressed by that 'title', the roles these men assumed, and the responsibilities they incurred, did not free them from the mundane business of gardening and hunting to provide food for their immediate family. Their status as personalised expressions of the world beyond was never generalised; they were not perceived as having power or superior knowledge beyond that entailed in the immediacy of their assigned role. Like everyone else in the community these men (and they were all men) were as likely as anyone else—indeed sometimes more likely, because their apparent prominence within the community could be interpreted as threatening local cohesion—to be challenged for perceived misdemeanours associated with customary expectations concerning exchange and sexual behaviour.

There were other domains, however, beyond those that made connections between individual men and the forms and functions of the outside world, where the emergence of 'modern' institutions was more powerfully expressed. These concerned local markets, school and church.

Market: Performing Modernity

Tuesday and Friday were 'market day' at Suabi. People began to arrive at the Suabi Corner market place from 7 am onward. Women brought small quantities of food for sale, displaying it on a sheet of plastic on the ground. If the river was in flood then people who lived south of the river could not attend. On different days different items predominated. Hibiscus leaves were regularly for sale while sweet potato, corn, lowland pitpit, pineapples and other food stuffs were sometimes present in quantity and at other times scarce. People bought from each other and, often, they bought foods that they themselves were offering for sale. Engagement with the social life afforded by the market was more fulfilling than the need to obtain particular kinds of food. Often, there were foods for immediate eating: edible bananas, cooked corn cobs, or buns made from flour and cooked in oil. These were popular, men buying them for smaller children or giving those children money so they could buy for themselves and learn early in life what money could do. Dried fish, a very few crayfish or, less often, cuts of wild pigs were sometimes available and these, though prices were higher, sold rapidly. Very rarely, when their own fresh supplies of vegetables were running low, men from the exploration base camp bought at the local market. For the most part, however, camp supplies arrived by helicopter at least once a week.

But locally grown or hunted foods were not the only items to appear at the market. Some people displayed small quantities of goods—torch batteries, soap, chicken stock cubes, packets of noodles and such like—that had been recently acquired at Kiunga, and one woman, a Highlander who had married in, regularly displayed clothing and, sometimes, bundles of dried tobacco. It seemed that everyone would like to operate their own trade store but that most lacked the means to do so. In fact, through the years 2011 to 2014 there was only one functioning trade store at Suabi. From time to time the man who ran it acquired moderately large supplies of goods—especially rice, cans of fish, cooking oil, sweet biscuits, Coca-Cola and other soft drinks, Digicel mobile phone FlexCards—which sold very rapidly.[10] Earlier attempts by others to operate trade stores—there had been many attempts over many years—had routinely failed, in part because the demands of close kin jeopardised potential profits, and in part, as various people told us, because the unremitting 'jealousy' of non-kin was discouraging.

The market was a clearly gendered space. Traders were almost exclusively women, whether produce was from gardens, forest or Kiunga. Men might wander the arena, a few cigarettes in hand, or a length of rubber for diving spears, subtly displaying these to potential buyers. But they did not sit and wait for custom. They were there as consumers, it seemed, rather than 'producers'. Admittedly, it was a man who ran the trade store, primarily men who dreamed of opening a trade store, or had done so in the past. It was men who at times sold small amounts of store-bought goods—rice, tinned fish, noodles—from their houses. And it was young men, too, who each week set up dart-boards at the market, selling nothing but hope and an opportunity to demonstrate skill. Unlike the women's stalls, however, these were all 'business', requiring serious capital to set up.

Throughout the course of the market, older youths and men spent much of their time playing darts, risking 10 toea for a single dart in the hope of a prize-winning hit. Older men gathered in groups to discuss concerns that might be aired at the close of the market when, nearly always, as sales ceased, it was time for a *tok save*—for an information talk. Several themes recurred time and time again: the need to bring community schoolteachers to Suabi; the need to clear the school grounds and rebuild

---

10   Sweetened biscuits, soft drinks and sugar were popular at Suabi by 2011, with soft drinks used to pacify crying children and sugar, when available, added to tea and sometimes used to make 'home brew'. In earlier years at Gwaimasi, sugar cane, which was thought of as a 'drink', and the uncooked larvae of wasps were the only sweet items on the menu. People did not eat honey produced by either native or introduced bees. It was too sweet and considered to be distasteful.

dilapidated classrooms and teachers' houses; the need to initiate work on a new community health centre and undertake airstrip maintenance; the need to clean open areas and tracks at each Corner; and the concern that unnamed 'school boys' were accessing 'drugs' (marijuana) and alcohol and damaging, or thieving from, community property. Other less regular themes included calls for employees to work at Juha, the failure of government to provide support to the Suabi community, and the threat that Mission Aviation Fellowship would cease flights if there was 'one more instance' of smuggling alcohol in personal baggage. And, on one very lengthy occasion, a man who held the position of 'Law and Order Committee' read all 40 items listed on a typed document that was headed '*Liklik los bilong Papua New Guinea wantaim ol panismen*' (the small laws of PNG and associated punishments). He first read them in Tok Pisin and then translated to Kubo. They included items such as '*dring bia long public ples*' (drinking beer in a public place) which was to incur either a PGK40 fine or a month in jail, '*tok nogut long narapela*' (speak badly of another person) which was to incur either a PGK300 fine or a year in jail, '*holim ol nogut video kaset na megesin*' (being in possession of pornographic videos or magazines) which was to incur either a fine of PGK2,000 or a year in jail, and '*man he selim pamuk meri*' (a man soliciting customers for a prostitute) which was to incur either a PGK800 fine or two years in jail. Many of the purported 'laws' were irrelevant to the Suabi community and, for those where there was potential for transgression, there was no way in which the declared 'punishment' could be imposed. Nor was anyone sure what should happen to the money if a fine had been put in place and paid. People seemed to assume that the money would rightfully belong to the person—the 'Law and Order Committee'—who imposed the fine.

The market place, therefore, provided a venue for public discussion of issues that had implications for the entire Suabi community and were conceptualised as 'modern'—as engaged with the world outside—in their content. It was not the place at which personal grievances between particular people or groups of people should be or were usually aired. This was made clear in a public notice displayed at the market in March 2014. The notice read:

> This toksave serves to inform the landowners of Soabi communities, the public servants and the General public that the public toksave regarding the personal issues has been closed as of 23rd March and no more.

> Any personal worries and issues that concerns you is not allow to toksave in the Market and Community.

The acceptable toksave in the market are if:

- Reasonable / or
- Genuine
- Urgent

Personal toksave as clossed to avoid confliction such as:

- Civil wars
- Gossiping
- Descriping and etc …

Note publish by the sensible guy and Acting Market Committee.

Personal issues were usually dealt with at local courts that, though on the surface formally structured in terms of understandings of the proceedings of government-run courts, were more faithful to earlier forms of resolving disputes. There were, however, two cases that did spill out into public debate at the market. On several occasions there was passionate discussion of inequalities between Corners as a result of school facilities, the community health centre, the ECPNG church and other mission buildings, and the market itself all being located at Suabi Corner. In consequence, speakers asserted, residents of other Corners were disadvantaged and, to some extent, treated as second-class members of the community. When this matter was raised, however, names were never mentioned. The other exception became more personal. It was thought by some people that several thousand kina of an infrastructure grant had been misappropriated. The problem was aired at a succession of markets, becoming more fraught each time until, finally, a 'suspect' was named and people came to blows. The altercation at the market then precipitated a court case where the matter was resolved (see Chapter 6).

At the market, people came together twice a week and demonstrated to themselves and to each other that they were, or aspired to be, 'modern' people. They were like people elsewhere—at Kiunga, at towns in the Highlands, at Port Moresby, perhaps even further afield. The potential that everyone could both sell goods—locally produced or acquired from distant towns—and buy what was offered for sale reinforced each person's participation in that 'modern' world. And, as well, the concerns that were usually aired at the close of the market were those that reached out to that world. The market, as performance, was aspirational.

But there were many other ways in which people—for the most part acting as individuals rather than as a collective—expressed themselves as 'modern' people. Older youths and men, if they had sufficient resources, wore boots, long trousers and dark glasses. They mimicked the safety clothing that featured at exploration camps.[11] Many younger women wore brassieres, and shorts beneath their skirts. Most mothers of infants used nappies rather than disposable soft leaves as absorbents. Some used 'baby carriers' that were supported across the chest in preference to carrying their infant in a string bag on their back, a very few experimented with very dilute formula as an alternative to breast milk, one sought a tubal ligation at the time she gave birth to her fourth child, and some young children were encouraged to learn to walk wearing shoes. A few older girls and young women purchased moderately expensive preparations with which they temporarily straightened their hair, while some used hair dyes. And nearly every household had an abundance of metal pots for cooking, purchased plates or bowls for serving food and spoons for eating. The implications of all these 'modern' accoutrements were, however, mixed. Some had been taken up for very practical reasons. Umbrellas were popular, especially for women who used them to shade infants that they were carrying. Men and youths who left the village to cut fronds of sago palms in preparation for thatching a roof now commonly wore gumboots and one heavy glove to make walking in the mud, where the palms grew, less unpleasant and to reduce the likelihood that they would be pierced by the abundant, long and sharp thorns of the palms. Straightening and dyeing hair—perhaps also the single case we observed of a young woman shaving her legs—presumably have aesthetic connotations. 'Modern' items associated with cooking and eating were hardly essential but were very convenient, though cooking pots increased the work of women because they blackened rapidly over the fire and much effort was expended in scraping them clean after every meal. Nappies, too, added to the work of the mothers of infants. But the new forms of clothing that men favoured—including heavy jackets that some had been issued when they worked in the mountains—were hardly appropriate to the high temperatures and humidity experienced at Suabi. They were statements of status, of membership within a new world of possibilities—of belonging

---

11　Martha Macintyre (2008: 183) comments that, in Papua New Guinea, 'in towns, in mining areas and around projects where young men have access to money' there are many who dress in ways that mimic 'modern' or 'global' styles and who do so as expressions of masculinity. We interpret analogous mimicry in dress by men at Suabi as an expression of a 'modern' status more than as an expression of masculinity.

to the category of those who wore such things. They were simultaneously, in their diversity of colours and styles, 'badges' of a person's individuality and, to this extent, placed that person apart from conventional relational and egalitarian practices.

School: Structuring Modernity

It was in 1987 that a community school, intended to provide the first six years of formal education, was first opened at Suabi. In that year anyone who had not previously attended school could enrol, with the result that some young men and women in their late teens and early twenties were both students and an immediately available workforce for building classrooms. Through the next ten to 15 years, government support to remote communities waxed and waned and payments to teachers were, at best, intermittent. The number of teachers posted to Suabi varied erratically and, often, those who did arrive were dissatisfied and sought placements elsewhere. By the mid-2000s the school was again functioning well. Now, however, trained local men taught students for their first three years—Elementary Prep and Elementary 1 and 2—and only Grades 3 to 6 were taught at the community school. At this time too, with exploration companies operating out of Suabi, there was more money available and a cohort of young men and women advanced to high school level at Kiunga or at several Highland towns.

The key community school teacher through this period was Eneka Jacob—a Kubo man. Eneka was committed to his work and largely responsible for the fact that a number of students achieved entry to high school. In 2009 he initiated a building program, upgrading what were then seriously dilapidated classrooms. But in September of that year, on a weekend fishing excursion with his half-brother, he was bitten by a death adder and died. Eneka's death was attributed to sorcery and, for this reason, other potential community school teachers—trained outsiders—refused to take up positions at Suabi. Eneka's building project was abandoned.[12]

---

12 By 2014, three burials at Suabi were marked by above-ground memorial shelters that displayed former possessions of the deceased men. These celebrated the lives and achievements of Digimo, the first Kubo pastor at Suabi who died in December 2005, the school teacher Eneka who died in September 2009, and Sosoaho who died in December 2011 when he was in his 80s—older than any other Kubo person—and who would have been about 25 years old when the patrol post was established at Nomad. The memorial structures were built at Owabi Corner in places of high visibility. By contrast, people who, after they died, were judged to have been sorcerers were buried with minimal ceremony at locations that were relatively distant from the usual places visited by other

In 2013 the community school was again opened. Two teachers came, though they departed before the school year had been completed. There was mutual dissatisfaction between teachers and community members. The former felt that the community was neither cooperative nor interested in their children's education, and submitted a negative report at the time they departed. The latter complained that the teachers had initiated illicit commercial ventures and paid unwanted attention to female students. In 2014, there was a long delay before a community school teacher arrived. The local school committee had failed to fulfil required financial obligations that would facilitate this process, and several teachers who had been posted to Suabi made alternative arrangements. Qiriwasi arrived on 2 April, in the ninth week of the first term of teaching. He was a Suki man but, in earlier years as a teacher at Honinabi, had married a Samo woman and they both came with their small child. Through the next two weeks community members devoted many hours to clearing the overgrown school grounds and restoring the two usable classrooms to some semblance of order. They had made only desultory contributions to this necessary work before the teacher actually arrived. Qiriwasi commenced enrolling students. He was conscious of the late start and initiated lessons on 17 April, during an official school holiday period. To the chagrin of some local people he also called students to school on 18 April—Good Friday: a 'public holiday', people said. At this time there were 54 students enrolled: 35 in Grade 4, 12 in Grade 5 and 7 in Grade 6. Qiriwasi felt that he could not handle Grade 3 as well. It was hard to imagine how, as sole teacher, he was going to provide for the educational needs of students in the three grades that he did accept.

The first elementary school teacher at Suabi, a Febi man whose family had long lived in the community, had sexual relations with some female students and was removed from his position by the local community. He left Suabi and, in 2008, was replaced by Taio, another local man, who had received one year of training. In 2011 Taio was joined by Dinosi and Okiset who had both attended a six-week training course and were, officially, eligible to teach only Elementary Prep. Taio was given the official status of 'headmaster'. In 2012, Okiset, who lived at Gweobi Corner, established a separate elementary school there to cater specifically

---

people. In all cases, burials were below ground, with the body placed on a low platform and protected from the earth by bark and timber walls and roof, and those who handled the body wore surgical gloves acquired from the health centre and, sometimes, face masks. The gloves were thrown into the grave before it was filled.

for the children of Febi-speakers. And, in 2013, Kabel, who had received a full year of training in Port Moresby, commenced teaching at a newly established Timaguibi school. Dinosi, Okiset and Kabel all worked under the supervision of Headmaster Taio, but Okiset had registered his school under the name 'Juha Elementary School'; in the understanding of distant bureaucrats, therefore, this was not part of the Suabi set of schools and was eligible to receive funds that were not channelled through the headmaster.

In 2014 there was great variation in the commitment of elementary school teachers. In the 12 teaching weeks that we were present, classes were held at Timaguibi in ten of those weeks. The first teaching week was devoted to preparing classrooms and school grounds and, in the last week, the teacher attended a professional workshop at Nomad. At Suabi Corner, the headmaster and his junior co-teacher ran morning classes in, respectively, only four and five weeks and at Gweobi there were classes in four weeks. The general lack of commitment was, in part, simply a reflection of the way in which Kubo and Febi people undertook many work-related tasks. Procrastination was commonplace. A sense of urgency or, more often, evidence that someone else had taken preliminary action—evidence, for example, that a teacher had actually arrived before commencing cleaning the school ground—was required to stimulate engagement. It was a reflection, too, of the fact that two of the teachers felt they were not adequate to the task and were receiving insufficient guidance from the headmaster. Officially they should teach only Elementary Prep but were called upon to contribute to higher grades. It was influenced by the fact that resources were minimal or non-existent; in part because, in some years, the Education Department failed to provide an expected grant, in part because when a grant was received relatively little of it was spent in the ways intended, and in part because at the close of each school year any remaining resources disappeared. And, finally, it was influenced by the fact that some of the teachers contributed their time and effort for long periods without receiving any financial remuneration. Dinosi and Okiset both taught in three years without being paid and Kabel taught in 2013 and the first term of 2014 without pay. Through this period only the headmaster received any pay, though this too ceased in the second term of 2013, when the provincial teaching authorities learned that he was spending much time in Kiunga when he should have been teaching.

In 2014 both Dinosi and Okiset were finally successful in negotiations with Education Department officers in Kiunga—they had tried on several earlier occasions—and received payments of, respectively, PGK14,900 and PGK12,000 for the previous three years of teaching. The headmaster was also relisted as an active teacher though, due to bureaucratic oversight, he was in the fortunate position that he too received approximately PGK12,000 despite the fact that he had previously been remunerated for times when he was thought to be at Suabi and purportedly teaching.[13]

Enrolments at the elementary schools were high in 2014, with 80 students in three grades at Suabi, 27 in two grades at Timaguibi and 15 in one grade at Gweobi. Ages ranged from about six years to early twenties, and 45 per cent of the students were female. However, a strong bias favouring males was evident at higher levels of education; only 33 per cent of the 54 students enrolled at the community school were female—there were no females in Grade 6—and only two of the 19 students (11 per cent) purportedly enrolled at high schools or technical colleges at distant towns were female.

Formal education has had a chequered history at Suabi. In the first ten years from 1987 a considerable number of students progressed beyond community school to reach higher grades—though seldom the highest grade—at high school. A few went further and undertook technical training in, for example, carpentry, welding or plumbing. But virtually all this cohort was male. There was a period thereafter—of five or more years—when relatively few Suabi students moved on to high school. A second phase of successes occurred in the mid-2000s and it was in this period, for the first time, that a number of females from Suabi attended high schools at Kiunga or in the Highlands—though none completed their studies and a few returned pregnant. That phase was short-lived, however, and by 2014, when once again there seemed to be a promise of local opportunities for education, those who attended community school were, in general, older than would be expected for the grade in which they were enrolled.

---

13 Dinosi and the headmaster supported their claims for salary by showing Education Department officials a variety of well-presented forms that provided spaces for enrolment details, weekly activity plans and daily lesson plans in 'language', 'cultural mathematics', 'me and my community' and 'me and my environment'. The forms suggested that much effort had been invested in planning the teaching year though, in fact, they were based upon forms provided during training and their rather well ordered, and multicoloured, appearance was because we had typed and printed them.

School: Learning to Be Modern

At Suabi, school structured people's learning experiences in ways that differed greatly from earlier times. And, together with church services, market days, timetabled clinics at the community health centre and scheduled radio contacts, it also structured people's organisation of time in new ways.[14]

At Gwaimasi, and neighbouring communities, in 1986–87 family sizes were small and children regularly accompanied parents or other kin to gardens and sago processing ventures where they observed, and in some ways participated, in the work that was done. As they became older and their interests turned in different directions, boys who wished to hunt larger game—wild pigs or cassowaries—took on an apprenticeship role with older, experienced hunters and girls developed competence at subsistence tasks and weaving string bags. There was little by way of formal instruction. Observation, listening, participation and experimentation were central to a child's education and what they learned was immediately relevant to their current and expected future needs. Their older kin, as incidental outcomes of their own activities, taught by directing attention to what the children needed to know and to ways in which they might accomplish what needed to be done. Children were enskilled rather than enculturated (Gibson 1979: 254; Ingold 2000: 22, 416; Pálsson 1994).

Education at schools is quite different. At Suabi, rote learning was common with, for example, pupils chanting vowel sounds, or words in English, in response to the teacher. Though, in fact, children developed more competence with both Tok Pisin and English through exposure to these languages in contexts other than school, the 'performance' of English in classrooms was seen as important. School education was valued by all, both adults and children. It was valued for the promise it held of a different and rewarding future, a future that could be accessed only if money was accessed; English, and the knowledge it seemed to represent, were understood to be ways in which this might be facilitated. As one man told the children at an end-of-year graduation event in 2011, 'muscles' were enough for getting food and building houses, but to get money they would have to use their 'minds' and school was the place to develop this capacity. Poor results in English told against graduation

---

14   Knauft (2002b) provides excellent accounts of the content and implications of classroom teaching, and of the scheduling of time, experienced by Gebusi people living at Nomad in 1998.

from elementary to community school and from the latter to high school. But few became fluent in speaking, much less reading and writing, this new language. High school reports for students often gave low grades for English. Some men, seeking fulltime positions with local companies—as helicopter loadmasters for example—failed in multiple attempts to pass the English component of an entry examination. At Suabi, those few with competence in written Tok Pisin or English were often called upon to draft letters for other people. Two men charged fees for providing this service.

What was taught at local schools had little relevance to the subsistence lifestyle that prevailed at Suabi.[15] The focus was on possibilities of future employment that would take those who graduated away from their home community. Only teaching and health work offered the chance to receive a regular income at Suabi and, at best, the number of positions would not reach eight. The returns from school education were not immediate, and their relevance was to a lifestyle that was far removed from that which most youngsters would eventually have to follow. Observation, listening, participation and experimentation were not central to school education. What was taught was, in large part, concerned with a world, and with ways of thinking, that were far beyond Suabi. It created expectations and desires but, in the context in which it was delivered, provided no built-in mechanisms by which those expectations and desires might be satisfied.

But school had more profound effects on life at Suabi. At the times when classes were run the school bell rang at 8 am. Most children who were attending assembled early. Few were late, though there were some who avoided school and found other ways to fill in the day. Classes at elementary school finished at midday, those at community school continued into the afternoon but, in both the morning and the afternoon, there were scheduled breaks. School structured time in an orderly fashion, creating a frame which patterned the lives of both the pupils and many of their kin. The markets held on Tuesday and Friday mornings, the regular church services held on Saturday for SDA adherents and on Sunday for ECPNG adherents, the clinics at the community health centre, the assigned days and times when Mission Aviation Fellowship or mission doctors were in radio contact, and advance knowledge that a plane was expected to reach

---

15 At the Kubo community at Testabi the head elementary school teacher instituted a weekly 'culture day' on which pupils were encouraged to attend dressed in 'traditional clothing'. There was nothing akin to this in the school program at Suabi.

Suabi at a particular time, all had the same effect. These were temporally fixed activities and people who wished to participate in them were obliged to pattern other activities around them. They had the outcome that many people were doing much the same thing at much the same time. This was in stark contrast to the flexibility, and family-level individuality, of activity scheduling seen at Gwaimasi in 1986–87.

Through the weeks when school was in session the children who attended could not travel with parents or others when these people went to gardens, sago processing sites or other places that were a few kilometres from the village. School, therefore, removed children from many of the activities that, in earlier years, had been central to their educational experiences. And those activities became less attractive to them. Many became less enthusiastic about travelling with their parents even at times when school was not in session. In the large Suabi community where, in the context of school classes, they were often with age mates, children increasingly associated with one another out of school hours—girls making and using fish traps, boys playing cards or mini-snooker, gender-specific groups swimming. A restlessness was evident, a restlessness that was exacerbated by the stories older youths and young men told of the excitement to be had at Kiunga or, especially, Highland towns, a restlessness that was often given expression by tensions within families and the frequency with which many teenaged children moved their sleeping arrangements between Corners and from house to house.

School: Expressing Modernity

At Suabi, there is little concern to capture anything of customary practices and understandings in the context of the formality of school education. Even at elementary school the emphasis is upon instruction in English and, to some extent, a disparagement of a child's natal language. A well attended graduation ceremony held in December 2011 gave emphasis to the 'modern'. With a single exception—a teacher who wore a cassowary feather headdress—neither adults nor children dressed in traditional clothing. The ceremony was organised to celebrate the achievements of elementary school pupils who, in 2012, would be advancing to a higher grade. But, regrettably, on the day, it was not known which children had qualified to graduate, because the headmaster was in Kiunga—his wife was due to give birth—and had not received or forwarded results of the end-of-year tests. Several hundred people gathered in the early afternoon. Entry points to the school grounds were decorated with streamers made from stripped palm fronds and flowers, and signs indicated where 'students' (girls and boys separated), 'parents' and 'professionals'—elementary school and university teachers, the Law and Order Committee, other community 'leaders'—should gather. The pupils assembled as three carefully spaced lines in front of the PNG flag, sang the national anthem—*O arise all you sons of this land,/Let us sing of our joy to be free,/Praising God and rejoicing to be/Papua New Guinea*—and moved to their appointed sitting places. The 'professionals' made speeches.[16] Then it was time for the 'entertainment'. The Owabi string band and the Suabi live band—featuring guitars and, in the latter case, a drum—presented items that they themselves had composed and practised through the previous weeks. The third and final item, listed on the hand-written program as a 'dramer', was titled 'Strit Mankis' (Street Boys) and presented by teenage girls from Timaguibi Corner. It was a rich performance, accompanied by disco music from a boom-box powered directly from a solar panel. Seven 'street boys' crouched in a tight circle, playing cards, smoking, drinking and beginning to stagger as they got increasingly drunk. They stood, formed a tight circle, and danced wildly. They did not speak. They did not look at the audience. They were self-engrossed. One 'boy' moved from the group to stand at two leafy branches—'trees'—and relieve himself. He was struck from behind by a *sanguma*—a 'magic man' or *hugai*— who he did not see, returned to the group and fell dead to the ground. His friends wept. A second group, girls and one older woman as pastor, silently watched the depraved behaviour of the 'boys', their disapproval evident. They quietly sang a hymn. The performance was over. The 'boys' had been punished but did not find redemption. It was probable that they would do the same again.

The Timaguibi girls had not learned of these matters at school. They had not created their drama in the context of school. Nor had they themselves been to the Highland towns where the behaviour they depicted was said to be common. They had drawn on the many stories that had been told and retold by both younger and older men who had visited Highland

---

16   We were invited to speak at the graduation ceremony. It was 11 years since we had lived among Kubo people and, on this visit, we had been present for only one month. We recounted a little of our own life journey, reminisced about earlier days living with Kubo, talked of people we had known, of being impressed by the number of children now living at Suabi, and of the fact that so many people spoke three languages when, in earlier years, they had spoken just one. We spoke of the value and advantages of each of those languages—the 'ground' or 'mother' language, Tok Pisin and English— and stressed, particularly, that they should hold onto the first, that it was a 'good language for a good place'. As we spoke, our words were translated. And when we finished, what was taken to be our key message was explained to the students: 'if you want money, learn English'.

towns and laughingly reported their own participation. The girls did not need words to convey their message. Their drama spoke of risks to their own community and, particularly, of the harm that could befall young men who, attracted by the pleasures of town life, abandoned responsible ways of living.

Church: Revealing Modernity

Church services, and other mission functions, were also occasions at which, with a few notable exceptions, past practices were seldom evident. ECPNG services took place nearly every Sunday, though at Christmas and Easter there were multiple services across four days and, once a month, a special communion at which participants were served 'tea'—warm water with powdered milk and sugar. Pastor Martin explained that ECPNG rules restricted communion to those who had been 'water baptised'. He himself would like the rule changed so that all 'believers' could receive the blessing of communion, and those who had been baptised but did not, in fact, believe would not be eligible to join in.

Sunday services ran from about 9 am to 11.30 am, attracting congregations of from 60 to 150 men, women and children. People sat on the floor, men to one side of the church, women on the other side. If the pastor was absent, the service was led by another man who had received some training at Bible school or, rarely, was foregone. There were hymns, prayer sessions, usually one of two 'testimonies' from members of the congregation, offerings of money, food or minor trade store items to support the mission and the pastor's family, a sermon and sometimes, at the close of the service, an open discussion of community concerns or of the need for practical help cleaning mission property. Martin put much effort into his sermons, was a natural teacher who drew on a comprehensive knowledge of the New Testament, a good performer able to elicit participation from those to whom he spoke. He alternated between Kubo, Tok Pisin and English.

In his sermons Martin spoke, variously, of the things that made a 'good home'—the presence of Jesus, a good wife, a good husband and a Christian marriage; of the need to 'fight evil' in the world; of the importance of coming, not just once but time and time again, to Jesus for a 'recharge'; of resisting temptation; of establishing a personal relationship with God before establishing good relations with others. At the close of each sermon he drew out his primary message: that, for example, the need to 'live in the light that was God' should be understood as the need to live a 'moral' and, hence, 'holy' life, while the 'darkness' where Satan was found was a place of 'mistakes' and 'sin'. Sometimes his focus was on matters he himself was grappling with, though at these times he preferred to open discussion to the congregation rather than reveal his own doubts and uncertainties. At one service, for example, he asked whether a Christian should eat of a pig that had been given as part of bridewealth, as a fine for adultery or as compensation for sorcery. If your unmarried son or daughter has been found to have had sex should you, as a Christian, contribute to the payment of the fine? If you are sick, or someone else is sick, and a 'witchdoctor' requests a monetary payment for offering a diagnosis or suggesting a cure, should a Christian accede to that demand? In these examples, Martin was abstracting from recent cases. His questions elicited prompt responses. A Christian should not do these things. But, to Martin, it was more complex. A pig, he suggested, is 'just a pig'. There should be nothing wrong with eating it. It is the context that should provide the clue. Only if the context is 'bad' is doing so wrong and not Christian. His challenge to his audience, and his own uncertainty, was to know where the cut-off was between 'good' and 'bad' in a continuum of possibilities.

It was matters concerning sorcery that Martin found most difficult to resolve. He himself had seen people die after being attacked by an assault sorcerer (*hugai*). He had seen people fall ill, and eventually die, in response to parcel sorcery (*bogei*). But he did not know how the old people knew these things, how they knew that the symptoms someone displayed were the result of sorcery of a particular kind or emanating from a particular source or how they, or others, knew how to ensorcel. He understood that assault sorcerers killed a person, stole organs, brought them back to life and sent them home where they would die for a second time. But this, Willie told us, contradicted his Christian understandings that only God could return someone to life. He accepted it as true—he had seen much evidence—but was bewildered by the apparent contradiction between customary Kubo beliefs and those that he had come to hold.

Martin was born in the mid-1970s. He attended community school at Suabi, moved to Nomad for technical training in carpentry, initiated training as an elementary school teacher and received some practical teaching experience at a Balimo school. In 1997, now back at Suabi, he organised the distribution of relief food that was provided by the Australian government during the course of a devastating El Niño-induced drought (Minnegal and Dwyer 2000a), and soon afterwards, encouraged by the local American missionary, trained as an ECPNG pastor. In the mid to late-2000s he became pastor at Suabi, and was eventually appointed to oversee all ECPNG matters in the area occupied by Kubo, Samo and Gebusi speakers. He travels quite often, providing support to church groups throughout this area (often walking a day or more each way to distant communities) or meeting with senior ECPNG representatives at Rumginae. He considers that his own skills within the church are not really those of a pastor, that he is not good working with the local community, talking to people in their houses about their life problems. His wife, Tabua, is better at this. His skill, he said, and the way he could best serve the church, was as a teacher travelling throughout the region that he oversaw and providing guidance with mission-related secular and sacred matters. For several years, at Suabi, Martin has run a Bible school with an enrolment of about ten young men and women who come, variously, from Kubo, Samo, Gebusi and Bedamuni language groups. He felt strongly that when pastors were telling Bible stories they should draw on both the content and style of the local stories of people. The people would then be better placed to understand the message the story was intended to convey. He said that when the first missionaries and pastors came they taught people to sing hymns, they preached about God but the people did not really understand, did not 'feel' the words. But that is changing. Now people are encouraged to 'sing their own feelings', not just words from a book.

As Easter approached in 2014, Pastor Martin encouraged women from each Corner to 'sing their own feelings'. We arrived late at the Sunday morning service and, as we crossed the grass to reach the church, heard women singing. We were transported 28 years into the past, for the way in which they sang was the way in which they had once sung at gatherings when men, painted and costumed, danced through the night. The sounds were subdued and mournful, emotionally charged, quite unlike the way in which hymns were usually performed. Their words were not those of the past—they sang '*Jesus is dead*' and not, as they might once have done, '*at the mouth of Bo, a scrub fowl is calling*', meaning that a particular man was calling at that place or had seen a scrub fowl at that place—but the style of performance was from that earlier time. And once again, as with the drama presented at the school graduation ceremony, it was teenage and older women who reached beyond now conventional practices and expectations to, in the one case, critique future threats to community well-being or, in the other case, conjure memories from a supposedly abandoned, even forbidden, past.

Expressing Community: The Past in the Present

At Gwaimasi, in 1986–87, wild pigs, cassowaries or large hauls of fish or cave bats were always shared by all those present at the village. When the food was cooked, people were called to each bring a plate. And everyone did so, young children, but not nursing infants, included. Everyone received essentially identical portions—a cut of meat, a portion of fat, skin, a section of entrails and so forth. The same principle patterned the distribution at feasts when visitors attended. Distribution of food was often a lengthy process, with observers drawing attention to any imbalances in the distribution. By 1995, however, a shift had occurred so that now feast food was often shared to 'groups' rather than individuals, with the identification of those groups based on *oobi* or village affiliation. But insistence on the equivalence of those shares with respect to size and composition persisted (Minnegal and Dwyer 1999: 70, 2007: 19).

At Suabi, in 2011–14, this principle of identity in sharing continued to be expressed at all feasts. There were many of these. They celebrated marriages and school graduations, resolved court cases, or marked the capture of a wild pig, departures, the conclusion of a working bee, and occasions of 'fellowship' for Mission adherents. If there were few people then everyone received a separate plate of food. If there were many, the distribution was to groups—to households if those who attended were from a single Corner, to wider assemblies if the feast brought people from different Corners or villages together. Rice and a 'soup' of vegetables, tinned meat (usually fish) and noodles were expected on such occasions. These were cooked at different households, then carried to the place where people gathered and the distribution occurred. Ten, 20 or even more pots of rice and soup might be provided. And, without fail, each plate—that of an individual or that of a group—received a portion from every one of those pots, though the content of each was effectively the same. What mattered, it seemed, was that all who attended received a share from each household that had cooked food and contributed to the occasion.

In earlier writing we argued that 'under the influence of monetisation Kubo … moved from an emphasis on equivalence in exchanges to a recognition of substitutability' in which the commonalities between things 'rather than their individual qualities … increased in importance' (Minnegal and Dwyer 1999: 70). By 1995, for example, people accepted that a pig could be exchanged for money and, when this occurred, it was the size and condition of the pig that determined its price rather than that it was a particular pig with a particular history. Increasingly, through the next ten to 15 years, money was depersonalised so that, by 2011, people were thoroughly familiar with the commensurability of different monetary tokens and freely exchanged money for desired goods. As Martin said when problematising 'good' and 'bad', and asking when it was appropriate or inappropriate to eat pig or contribute to buying pig, 'a pig is just a pig'. There was something deeper, he was suggesting, that should guide a Christian's decision. But, at mission feasts, where his own role was always prominent, Martin did not behave as though 'a pot of rice is just a pot of rice'. At feasts, more than on any other occasion, there remained an emphasis on equivalence—on publicly acknowledging the contributions of particular individuals or particular households—in the exchanges that occurred.

Social change is never absolute. The past is never put entirely aside. Constructions of identity run deep. And practices that fulfilled some needs in the past may continue to have relevance in the present. Such practices may be mundane or sacred. Among Kubo people, through the 1990s, the number of people who smoked declined notably and in our own later visits to Gwaimasi there was no one, other than us, who smoked. But this reduction had a cost, for the exchange of smoke—where one man drew smoke into a bamboo pipe and gave that smoke to another man—was a usual, and appreciated, greeting ceremony. Men 'ate smoke' together.[17] By 2011 there had been a resurgence of smoking. It was common among men who had spent time away, employed by exploration companies, and among older youths who had been to schools in the Highlands. Some men grew their own tobacco, others bought trade tobacco and a few sometimes bought pre-rolled cigarettes. Men returning from Kiunga might carry a newspaper and sell pages that were then used to roll cigarettes. And, often, two men—especially a host and visitor—would express their enjoyment of each other's company by 'eating smoke' together.

More striking, however, was a resurgence of customary ways in which to treat people who were sick and may have been ensorcelled. In the first few months of 2014, three curing dances were held; the first on behalf

---

17 Knauft (1987: 75–80) wrote that for Gebusi the sharing of tobacco was 'the *sine qua non* of male social life'. Tobacco use, he said, was brought to its fullest expression at ceremonial gatherings, and in ritual fights was consciously used to forestall anger.

of a woman who had been sick for some time, the second and third, a month apart, on behalf of a man who had been attacked by an 'assault warrior'—a *hugai* (Fig. 4.4). A sick child was included in the second ceremony; a sibling had recently died and the parents, anxiously, were seeking to protect the living, to offset the likelihood that this other child would be subject to the same spiritual attack. At each ceremony a spirit medium from an outside community officiated and one or two men danced through the night to the beat of a drum. The medium called upon benign spirits to dispel others that caused harm. The dancers attracted their attention. The chosen venues were distant from the ECPNG church; some people were deeply concerned that the performances were not Christian and felt they should not have taken place.

Figure 4.4: Curing dance at Suabi.
Source: Photograph by Peter D. Dwyer, 2012.

There had been three curing dances at Gwaimasi in 1986–87, one following a séance, and at intercommunity feasts all-night dances were an integral part of events. But we had seen none of these on later visits. At Suabi, as at other communities to the south, traditional dress and dance had become relegated to daytime performances at Independence

Day celebrations. These were occasions for performing 'cultural' identity, competing with groups from other places to put on the most colourful and 'authentic' show, often for prize money. Negotiating relations with those others, or with the spirit world, were no longer the point of the exercise. In late 2012, however, for the first time in many years, men again danced through the night at Owabi during a major feast that drew together visitors from all Kubo communities, as well as Febi, Samo, Bedamuni and others from further afield. More dances were held during smaller feasts the following year. There remained an element of 'performance' about these events, but that performance was now for each other, it was not primarily for consumption by outsiders.

Circulating Money

Through the past 20 years money, and the desire for money, have become more important to Kubo and Febi people and have increasingly shaped both practice and understandings. But the amount of money entering the community is not great. Expectations of future windfalls greatly exceed the current reality. In four-and-a-half months from mid-December 2013 to May 2014 a minimum of PGK250,000 reached the community as cash, with about one half of this paid as wages to men and women employed on a casual basis at the local base camp or as labourers on seismic lines in the vicinity of Juha (Table 4.2). While we may have been unaware of some payments to the community, gossip about money is commonplace and it is unlikely that the maximum reached PGK300,000. Three elementary school teachers and the community health worker received, in total, approximately PGK41,000 and small contributions were made to the local ECPNG pastor and the man who acted as agent for Mission Aviation Fellowship. Grants from government contributed another PGK80,000, most of which was spent on local employment. Talisman made a one-off payment for rent and environmental damage associated with the camp established near the airstrip. And, finally, some cash payments were received from us and, probably, as remittances from the two men who held permanent positions—one as a community health worker, the other as a loadmaster for Pacific Helicopters—beyond Suabi during that period.[18] The estimated total is equivalent to PGK75.9

18   Our own primary cash inputs to the Suabi community were in 2012 when we purchased and paid for upgrading a recently built house. We often reciprocated assistance by way of gifts and, in this

per person per month, roughly the same as our estimate of per person annual income at Gwaimasi in 1986–87. At Suabi, however, in contrast to Gwaimasi, the distribution of money was markedly uneven across community members with a very marked bias against women.

Table 4.2: Money entering the Suabi community between 19 December 2013 and 5 May 2014.

Source of money	Estimated amount (PGK)
Grants	80,000

Local employment*
  Gas and Oil
  Education & Health
  Mission
	125,000
	41,000
	500

Rent Compensation	6,400 4,215
Other*	4,000
Total	261,115

Source: Authors' work.

Note: 'Other' combines remittances and our own payments; * = estimate.

Moreover, there is little consistency from year to year in the amounts of money, or the sources of money, entering the Suabi community. In the years 2012 and 2013, the only grant received at Suabi was of a few thousand kina to support the elementary school. The money received by elementary school teachers in 2014 covered the previous three years of employment. Employment of local people by exploration companies was virtually non-existent in the years 2009 to 2012, and was much higher in the closing months of 2013 than in the early months of 2014.[19]

Substantial amounts of money were spent buying trade store goods at Kiunga and shipping them by plane to Suabi. Food items predominated, with rice accounting for an outlay of at least PGK12,000 through the four-and-a-half-month sample period.[20] Most rice and other food items were

way, reduced both the frequency of requests for money to, for example, support the school fees or transport costs of particular individuals and the likelihood that monetary contributions to one person would elicit 'jealous' responses from others.

19   Late in 2011 we spent one month at Suabi. For some time before that, monetary returns to the community had been very low. Some people hoped that our presence would compensate for this but our needs were relatively modest. On one occasion some men complained that we had not provided long-term employment for any men. Several women responded vigorously to the effect that while it was true that our contributions did not match those of visiting petroleum companies, at least we dispersed payments to men, women and children in ways that the companies did not.

20   Our estimate of the outlay on rice is based on counting bales of rice—20 kg per bale—offloaded from planes, and aggregates retail and freight charges.

resold within the community with a small mark-up in retail price, though in cases where the original buyer had personally made the purchase in Kiunga—some orders were placed by mobile telephone—it was rare that the mark-up was sufficient to cover the buyer's airfares. The primary outlay was, however, air transport to and from Kiunga, Highland towns and, in a few cases, Port Moresby. In four-and-a-half months, the 143 residents at Suabi Corner—19 per cent of the Suabi population—spent a minimum of PGK23,910 on airfares to and from Kiunga and Mount Hagen, plus an additional PGK10,000 on two charter flights. The estimate excludes additional charges for excess freight. For the community as a whole, more than one third of the money entering Suabi through that period was spent on travel. Other notable expenditures that removed money from the local community were purchases of mobile phone Flex Cards and contributions to the needs of youths and young men who left Suabi to further their education.[21]

Though much of the money that entered the community was rapidly spent, a considerable amount was held—preferably, but not always successfully, secretly—with the intention of servicing a variety of social needs internal to the community. On a few occasions, when money that had been held secretly was lost, the amount entailed was discussed openly though often with some scepticism. Yameka, for example, carried his money with him when he left the village with his family to work in gardens or cut firewood. One day, returning home in the late afternoon, his small sons tipped the canoe over. Yameka's possessions, including several hundred kina that he had carried, were lost. Fiabo also lost perhaps as much as PGK1,000 when, in daylight hours, a mentally-handicapped

---

21 Mobile phones were not available at Suabi until 2011 when Digicel built towers at Nomad and Mougulu. The former, though closer to Suabi, was initially faulty and could not be accessed from Suabi until late in 2013. The latter, 27 km to the southeast of Suabi, was more reliable, though reception could be assured only at night and only from a high hill that entailed a two-hour walk and an overnight stay. In 2013 the Nomad tower was upgraded, reception improved and people made calls just on dark from the top of a ridge at Owabi Corner. Recharging telephone batteries was problematic. At first, the ECPNG Mission provided this service via a solar-charged battery and at a cost of PGK1 per recharge. In 2012–13 we provided this service to many people at no cost, but the demands on our own limited solar power meant that the practice could not be sustained. In 2013–14, some people had access to small, portable LED-light lamps that had been provided by ExxonMobil and could be used to charge phones. Others prevailed upon workers at the base camp to recharge their phones. Camp managers turned a blind eye to this use of company resources, though they were sometimes irritated when they felt the demand was excessive or too many unauthorised people were moving through the camp. In the absence of the preceding possibilities, some people had devised their own rechargers using a series of D-cell (1.5 volts) torch batteries.

man burned down his house.[22] Two primary categories of social needs entailed exchanges of money—locally-convened court cases and marriages. Both had the effect of redistributing money across the community, and both revealed inflationary effects across time and considerable ambiguity with respect to both value and sustainability. On the last count, on the one hand, people were conscious of the fact that monetary returns to the community varied through time and increases to fines or bridewealth in the present might not be able to be met in the future and, on the other, most were convinced that in the near future they would be in receipt of huge benefits from the PNG LNG Project.

Court Cases

Local court cases are discussed in Chapter 6, but one example may suffice to illustrate ambiguities with respect to appropriate value.

In 2012, Dougal initiated an affair with a young woman who had a small child and was considered to be married to that child's father, Asnah. Elena lived in Asnah's father's house, though Asnah was absent at school in the Highlands and some people knew that he had no intention of committing to the marriage. Elena became pregnant to Dougal and they lived together as a married couple. In December, Asnah returned to Suabi and, though he was Dougal's clan brother, called for him to face a court. After two hours hearing and discussing 'evidence' Dougal was fined PGK5,000. The fault at issue was the harm that Dougal had done to Asnah by marrying his wife. There was no discussion of arrangements for the child of Asnah and Elena; that was for the future. Dougal responded to the announced fine by saying that he did not have PGK5,000, and could not pay that amount. Without further discussion, the men who had imposed the fine immediately revised it to PGK1,000. Dougal accepted this. Now, however, people who had attended the court case—including one man who was married to Asnah's sister—handed over money to 'help' Dougal. The amount contributed was PGK935. This was judged to be sufficient,

---

22 Fiabo's house was burned down in December 2011. This was the third house burned by Orry over a period of four or five years. It was understood that in some way—he was described in Tok Pisin as *longlong* (stupid or crazy)—Orry was not responsible for his own actions. People did not know how to handle him. Some older youths suggested he should be killed and quite strong efforts had to be made to discourage them from doing so. After Fiabo's house was burned, police in Kiunga were contacted by phone. They were unwilling to come to Suabi, and suggested that Orry be brought to Kiunga where they would question him. Two weeks later Orry was encouraged to accompany two men who were walking to Kiunga. He was not delivered to the police, but has lived there ever since.

and given to Asnah. The man who had run the court case, the locally-appointed Law and Order Committee, now spoke strongly to Dougal. Different people had helped Dougal by contributing different amounts of money. Some had contributed less than PGK20. Dougal could forget these; he need not reciprocate them. But others had contributed PGK20, PGK50 or even PGK100. These should not be forgotten and should be repaid in the future. The court case was now over. The key protagonists shook hands.

At Suabi, many people were fined for one reason or another. But there was no consistency in the amount of either an initially suggested or finally accepted fine. The initial suggestion was influenced by thoughts of how much money the 'erring' person was likely to currently hold or be able to access in the future through employment or networks. There was no sense that a particular category of 'wrong' merited a fine of a particular amount. Indeed, the list of 40 *Liklik los bilong Papua New Guinea* was unhelpful for it had nothing to say about the kinds of 'wrongs' that were aired most often at village courts—sexual misdemeanours and accusations of sorcery. But the court case described above revealed more than ambiguities with respect to value. At one level, the structure of the court mimicked people's understanding of procedures followed at government level. One man acted as 'magistrate', 'evidence' was heard and assessed and, if warranted, a penalty was imposed. The epistemological underpinnings of the structure were categorical. At another level, however, pre-existing relational imperatives were never foregone. Anyone, whether male or female, was free to present and interpret 'evidence' and, with rare exceptions, was not silenced by the man acting as 'magistrate'. More significantly, however, once the amount of a fine had been settled—a settlement to which the 'offender' could contribute—many people asserted their existing or desired connections with the guilty party by contributing to the amount of money required. They were seen to contribute though, often, they concealed the amount. Some contributed despite the fact that their kinship affiliations were with the aggrieved party, and others despite the fact that as committed Christians they strongly disapproved of the offence now judged to have been committed. The resolution of the court case was understood to have righted the wrong and, though other interpretations might emerge in the future, it was necessary to reaffirm relationships and the sense of generosity that bound the community. This was an opportunity, too, to create situations of perceived debt that would both ensure ongoing reciprocity and reduce the likelihood of ensorcelment.

Marriage Exchanges

To 1986–87, most marriages among Kubo entailed the immediate exchange of 'sisters' by the soon-to-be husbands. Where immediate exchange had not occurred because a couple eloped, or because one of the men was without an eligible and willing 'sister', a variety of strategies were employed to create balance at some time in the future. A relatively common solution to the problem was for the couple to give a daughter to the woman's parents, as a 'sister' whom one of her brothers might exchange for a wife in the future. At that time, there was no suggestion of adopting the widespread PNG practice of bridewealth. But this was soon to change. In early 1988, Daledi and Gufu eloped. They slept overnight at a small garden house south of Gwaimasi and the next morning built a raft, travelled down the Strickland River and temporarily relocated to the longhouse at Gugwuasu. Daledi's only sister was already married and he was without an alternative. Gufu's classificatory father demanded monetary compensation, and by 1991 had received PGK480. Increasingly, through the 1990s, monetary payments were demanded or offered where immediate exchange could not be arranged, and by the end of that decade payments in the order of PGK500 had become commonplace. By the mid-2000s, with exploration companies in residence at Suabi, many men employed, more knowledge of Highland practices, and a few outsiders marrying Kubo women, giving bridewealth became a common alternative to sibling-exchange and amounts given rapidly inflated. By 2011, both were presented as acceptable Kubo and Febi practices. Some people preferred one practice; some preferred the other.

Increased access to money was one reason for the emergence of bridewealth as a legitimate basis for negotiating a marriage, but demographic changes were perhaps of greater importance. Men and women now lived longer. In 1986–87, in the communities around Gwaimasi, we knew few people in their forties. When Sinage and Tobu, Uhabo and Monu were negotiating their marriages in the mid-70s, their parents were already dead. By the time we returned to Suabi in 2011, however, parents were very likely to be alive when their sons or daughters married. Being elderly, however, they were less likely to have been employed than their offspring. They tended to favour bridewealth because, in this case, they would receive part of the payment in return for the effort they had contributed to raising their daughter. They argued in favour of bridewealth as a mechanism for personally accessing money.[23]

In the years 2011 to 2014 negotiated bridewealth was of the order of PGK4,000 to PGK5,000 among Kubo residents of Suabi, but considerably higher among Febi residents. In one case, a young Gebusi woman who visited her sister and brother-in-law while they attended Bible school at Suabi agreed to marry a Febi man. Arrangements were negotiated with her parents at Yehebi by the mission radio. The parents opted for bridewealth; they had no son eligible for marriage. The bridewealth was set at PGK8,000 though the money was not immediately available and was to be paid at some time in the future. Horace was also a Febi man. He had two wives but had not provided an exchange 'sister' or bridewealth for either of them. His case was taken to court where it was determined that, for the first of those wives, he should pay PGK10,000 together with two pigs each valued at PGK1,000 plus some 'coins' to produce a total equivalent to PGK12,400. Horace did not have that amount of money available, so again, though the price was set, payment was delayed to a future time. No decision was reached with respect to his second wife.

The inflation in bridewealth observed at Suabi is, in large measure, driven by the understandings, expectations and desires of Febi residents. They, more than Kubo residents, are confident of future monetary benefits from the gas wells at Juha. They have a long history of acting as middle-men in trade networks between lowland and Highland language groups (Goldman 2009: 2–38), have lengthy exposure to and, on the basis of some marriages contracted with northern neighbours (Bogaia, Duna and Huli), experience of bridewealth as an alternative to exchange of sisters. Further, in the past decade a number of Febi men have sought to establish affinal connections with Highlanders by seeking brides from

23   Jorgensen (1993: 69) discussed changing marriage practices among Telefol in response to the government station established at Telefolmin in the 1950s, the Ok Tedi mine from the 1980s, and associated monetisation. He noted the 'extractive intent' of parents demanding bridewealth from outsider men who sought to marry their daughters. In addition, however, by the 1970s, the inclusion of money in bridewealth transactions had an effect that differed markedly from the situation observed at Suabi. In customary Telefol practice, the parents of a bride-to-be had the initiative in selecting a suitable candidate spouse. The preferred arrangement entailed village endogamy. Increasingly, this broke down as young women actively pursued marriages with outsiders. The outcome was to reduce the network of community members with a stake in the exchanges entailed by the marriage, whereas at Suabi the shift to bridewealth has led to expansion in the network of those interested in the outcome of marriage negotiations.

that region. Indeed, some Highlanders have encouraged the practice as a means of establishing their own rights to future benefits from Juha. In one case, the councillor for a small Febi community used the entirety of a PGK20,000 government grant—officially made to facilitate clearing land for an airstrip—as the down payment on a Highland bride. In 2014, several men at Suabi indicated that they had agreed to pay PGK20,000 or PGK30,000 for such brides. To this time, however, the only men who might have this kind of money available are the few who are nominated recipients of outside grants and are willing to redirect funds received to satisfy personal desires.

Marriage and Adoption

In 2014, the Suabi population included 22 widows, five divorced women, seven men with two wives each and many children who had been adopted. These statistics are underlain by changes in the organisation of both marriages and families. There were, of course, widows, divorced women, men with multiple wives and adopted children in earlier years. But the underlying reasons are altering. In the 1980s and 1990s, polygamous marriages usually took place when a women's husband died and she remarried a man who was classificatory brother to her deceased husband. Indeed, if a man had an affair with his brother's wife, he might be rebuked by being reminded that his brother was 'not dead yet'; he should be patient, knowing that she would become his wife if and when his brother died. Missionaries discouraged multiple marriages but, we were told, did not object to remarriage after a spouse had died. In the early 1980s, when Gisio's first husband died, she was keen to marry his brother but had recently 'become Christian' and sought approval from the local missionary before doing so. Until a decade or so ago, therefore, it was only older men who might have two wives. By 2014, however, at least four of the second marriages—none were leviratic—had occurred within the previous six or seven years and there were other men, quite recently married, who openly discussed their desire to find second wives. For both men and women there are now fewer constraints on sexual relations than had been the case earlier and this, combined with more people living in close proximity and more men likely to be absent for lengthy periods, has enhanced the likelihood of both adulterous and premarital liaisons. Three case histories are summarised in Figure 4.5.

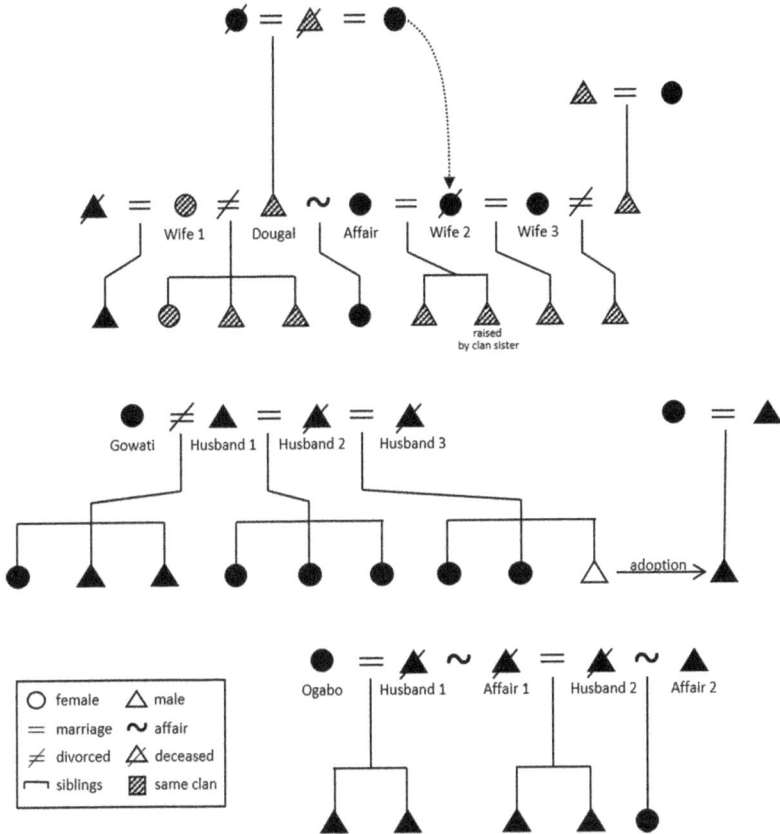

Figure 4.5: Marriage arrangements—three case histories.
Source: Authors.

Dougal married for the first time in the early 1990s, when he eloped with Gwaitidia, a young widow from his own clan. His behaviour was the subject of much disapproving gossip throughout Kubo territory. Gwaitidia's first husband had died shortly before she gave birth to a son. In late 2000 or early 2001, Dougal initiated an affair with a recently widowed woman, with the outcomes that his wife left him and a daughter was born to the widow. During this period Dougal's father died, and soon afterwards he married his father's widow, Beso. It was rumoured that Dougal and Beso were sexual partners before the death of Dougal's father. In 2008, at the birth of her second child to Dougal, Beso haemorrhaged and, after ten days, died. Dougal now invited Gwaitidia to re-establish their earlier relationship. She refused and, in early 2012, he initiated the relationship with Elena which led to the court case and PGK1,000 fine discussed above. All Dougal's marriages were initiated as elopements—one to a clan sister, one to his father's wife and the third to his clan brother's wife. None had received prior community approval and all remain subjects of unresolved tensions.

Gowati's first marriage was in January 1987. She was about 13 years old and, at marriage, relocated from Gwaimasi to Suabi. She had three children with her first husband, initiated an affair with a widower, married him and had three more children. When her second husband died she initiated an affair with a third man, who died in 2011 when she was pregnant with his third child. After his death, her third husband was diagnosed as the sorcerer responsible for some earlier deaths in the community. Because she did not want her last child to know that his father had been a sorcerer, she gave him to another couple. Her three husbands were from different clans. Her first marriage was balanced by an immediate exchange of a young woman from Suabi to Gwaimasi, but her later marriages were not reciprocated and remained subject to negotiation. One daughter to her second husband was 'given' to her Gwaimasi kin as a future exchange 'sister' but, by 2014, some of those people said that they now had enough 'sisters' of their own. They wanted to return the girl and receive a monetary payment instead.

Ogabo's first husband was killed in 1991 when, during a storm, a tree fell onto the forest shelter where he and some companions were sleeping while on their way back to Suabi from a feast. A few years later, after an affair with a young man, she married Hami, a widower with whom she had two sons before, in May 2000, he too died. Ogabo's marriage to Hami was not leviratic, with the outcome, after several years of dispute, that her first two children were eventually claimed by, and relocated to live with, their father's 'brother'. Soon after Hami died, Ogabo gave birth to a daughter as the outcome of an affair with Dougal. That child's paternity is well known but is not acknowledged by Dougal and is not a topic of public conversation. She is ambiguously placed with respect to both living arrangements and clan affiliation.

Only two of the 11 male–female relationships—the first marriages of Gowati and Ogabo—highlighted in these case histories conform to Kubo ideals and expectations. The majority are, as yet, unresolved with respect to either an exchange that would balance the marriage, a monetary payment that would stand in its stead or, where an affair led to the birth of a child, serve as compensation to those who were cuckolded.

Dougal, Gowati and Ogabo have been able to act in the ways they did because pre-existing sanctions—communal pressures to conform—have become less stringent and less effective, and because most people now feel that where relationships have been jeopardised by non-customary behaviour they can be eventually restored, at least on the surface, by an exchange of money. The case histories also illustrate the fact that there is now less concern to meet the expectations of immediacy than prevailed in the past.[24]

In the years 2011 to 2014 there were ten marriages at Suabi. One was constructed as balancing a marriage that had occurred a year earlier, in 2012. None of the others was balanced by providing an exchange 'sister' and, to May 2014, none of these had been reciprocated by payment of bridewealth. In all but three cases both the man and the woman were residents of Suabi. The three exceptions were marriages with, respectively, Samo, Gebusi and Konai women, one of whom was a widow who became second wife to a Suabi man. In one of these three cases the young woman was pregnant at the time of the marriage, though this was not known to her husband-to-be. When she gave birth her husband threatened to kill the infant by throwing him into the river. He did not want to support another man's child, though he did not contemplate divorcing his wife. His sister and brother attempted to dissuade him, arguing that the male child would be of his 'clan' and, hence, secure the future of that clan. He did not change his position, so his older sister took the child and gave it to her married daughter to raise.

Adoption has always been common among Kubo. Lifespans were relatively short, and many children lost one or both parents when they were young. When this happened they were immediately accommodated within other families, usually those of close kin—commonly that of their father's brother who, if it was the child's father that had died, may well have married his brother's wife. In the then-prevailing circumstances of a fragile demography, with high death rates of both adults and children, rapid and unambiguous adoption of orphaned children

---

24 The shift toward bridewealth in negotiating marriages, and the increased range of people—particularly the bride's brothers and parents—seeking a share of that money, placed greater pressure on young women with respect to marital choices. At the same time, however, the freedom of women had been enhanced with respect to acceptance of births that occurred outside of marriage and in reduced expectations that widows would remarry. Two long-term widows and one recent widow were the only adult members of one Suabi household. An earlier recognition that a young woman could veto the suggestion that she marry a particular man still obtained in the years 2011–14.

helped ensure that future marriage partners would be available for the adopter's own biological children (Minnegal and Dwyer 2006: 121–22). The kidnapping of children in association with raids in which adults were killed served the same purpose. By contrast, however, children born to unmarried women were frequently killed at birth—a practice continuing into at least the late 1980s.[25] The rationale in such cases was, in part, that the marriage prospects for a young unmarried woman with a child were unfavourable and, in part, that the child had no acceptable status as a member of an *oobi*.

By 2011, the frequency of births to unmarried Kubo and Febi women had increased substantially but, though sometimes threatened, infanticide was no longer practised. In some cases the unmarried mother raised the child herself, but more often, after weaning, it was cared for, and effectively adopted, by the young woman's parents. Adoption was very common. As before, orphaned children were adopted. In one case, noted above, adoption was intended to remove a child from learning that his deceased father was a sorcerer—a decision legitimised, in part, by the fact that the father had died before the child was born, so 'had not seen' the child and thus made it his own. And married couples whose families included more sons than daughters often sought to balance sex ratio by adoption (compare Demian 2006). In 2013, a man who had several sons paid PGK200 for the female child born of an affair by a divorced woman. By this time, however, adults lived longer and the survival rate of children was much higher than had been the case 20 or 30 years earlier. Family sizes had increased dramatically. And, further, many first marriages occurred after the young man and woman had initiated sexual relations. In a statistical sense there was less need to pattern adoptions with regard to future opportunities for sibling-exchange marriage. The increasing shift to giving bridewealth reinforced this change. What has emerged, however, is an emphasis on clan interests rather than family interests, an emphasis that has the effect that social reproduction and biological reproduction are less tightly coupled than they were in the past (Minnegal and

---

25   In the past, not all infants born to unwed mothers were killed at birth. One woman, who we have known since 1987, was born out of wedlock in the early 1970s. She was raised by her father's parents. Though for many years she was spoken of as being sister to her father, he now acknowledges her, with some pride, as his daughter. The same man has adopted a child born out of wedlock to his son and, again with pleasure, told us that she calls him *ade*—father.

Dwyer 2006). Adoption provides opportunities to strengthen a clan, directly in a numerical sense where the adoptee is male and indirectly through future affinal connections where the adoptee is female.

Finally, the reduced concern with ensuring immediate exchange in patterning marriages, combined with the fact that life spans were extended, influenced the living arrangements of newly married couples and the personal relationships that these facilitated. The earlier ideal, and common practice, was that men who had exchanged 'sisters' lived together in the same house, in neighbouring houses or at least in the same small community. The men would work gardens together or hunt and fish together and their wives would often collaborate in sago processing ventures. Through their earliest years, the children of each couple would be closely associated with their mother's brother—their *babo*—and, if the child was a boy, it was this man who would be his primary sponsor at the time he was initiated. The relationship was close and long-lasting. Preferred living arrangements ensured these outcomes.

But where the exchange of 'sisters' is greatly delayed, or has been replaced by payment of bridewealth, these living arrangements are less easy to put in place. At Suabi, it was common that newly married couples lived with, or alongside, the parents or older brothers of the husband. Less often they would live with, or alongside, the wife's parents. Clusters of houses were often those of extended families, unlike the mixed-*oobi* pattern that characterised communities like Gwaimasi in the past. A concern to support ageing parents, combined with the expectations and subtle demands of those parents, reinforced this emerging spatial organisation. The *babo* relationship—mother's brother, sister's son (MB/ZS)—remained important, but was less often given expression in everyday interactions, and less likely to being focused on a particular MB or ZS out of all those who might be similarly classed. And where a newly married man chose to live with his parents there was the potential that his wife would be disadvantaged—imagining, sometimes with good reason, that her autonomy was jeopardised by the demands of an ever-present mother-in-law. This was Alice's experience in the months before she and Jeff fought.

Three Decades of Change

In the years to 1999, at Kubo communities on the Strickland River, we observed an ongoing erosion of the relational imperatives that informed and patterned people's lives, and the infiltration of categorical ways of knowing the world. Women, pigs and land were being progressively revalued as 'things' in their own right—as instances within broader categories—rather than, as had been the case, according to the relational connections within which they were embedded. We have argued that the logic of money—as trope—was extrapolated to these domains of people's lives (see Chapter 2). That extrapolation was, in large part, figurative. No one was making the connection that we have made and, by and large, the changes observed, in discourse and in practice, concerned issues of exchange that were central to ways in which identity was understood and expressed.

Emerging in parallel with these shifts in the way the world was known was an increasing emphasis on the individual that was expressed in the size, structure and privacy of houses and the material items that a person possessed. Here, money facilitated the changes; it did not provide the logic upon which those changes were built. Money provided access to new kinds of material goods, and that access was differentially distributed. There had been no such differential when people could take what they needed or wanted from the forest. A desire to hold what you had infiltrated a pre-existing ethos of sharing all that was available.

By 2011–14 these changes had consolidated. There was now little ambiguity about exchanging women or pigs for money but, though money might be expected for the use of land by outsiders, no one yet spoke of the value of land *per se* in monetary terms. The way in which the external and internal structure of houses enhanced privacy for all occupants, or for individuals within that collective, was heightened relative to earlier years. But, of greater significance, and as revealed throughout this chapter, categorical ways of knowing were now manifest in day-to-day living— in, for example, the way time was structured, the ways people dressed to reflect desired status, the surface formality of court cases and, so frequently, in foregoing the relational concomitants of a person's name or kinship by addressing them, or referring to them, according to a 'modern' assigned role—Medic (Community Health Worker), Tisa (Teacher), Het Tisa (Head Teacher), Kaunsil (Village Councillor), Agent, Pastor and so forth. Their place in the 'modern' world—the category which they represented—and, by implication, the responsibilities that inhered in that position, was foregrounded.

Through a period of 28 years Kubo and Febi people have come to know the world in new ways. At the same time, however, the 'things' that they know have also changed. It is these changes that will be emphasised in the next chapters as we show how a deep concern to facilitate access to future financial benefits has set the scene for emergence of new social forms.[26]

---

26   Much of what we have described in this chapter differs in significant ways from patterns observed by Bruce Knauft at Nomad. Knauft conducted detailed ethnographic work with Gebusi, who live to the immediate south of Nomad (see Figure 1.3), in 1980–82 and subsequent work, which emphasised change, in 1998, 2008 and 2013 (Knauft 1985, 2002b, 2007, 2010, 2011, 2016). By 1998, the community with which Knauft had first worked had relocated to the government station at Nomad where, by and large, the resident Gebusi people were compliant with the ordered and bureaucratic impositions of policing, schooling and Christian worship that prevailed there (Knauft 2002b). By 2013, the airstrip at Nomad was no longer operating, and government staff had departed. Local people were reorienting their attention to the bush (Knauft 2016). Money was scarce, the local market little used, communal sports largely abandoned, and school attendance low. There was little theft, people no longer wanted wristwatches or bothered to keep track of hours in the day. All, it seems, still identified as Christian, though some people were beginning to explore alternative versions of Christianity and enthusiasm for church services had declined. The expectations of bridewealth that emerged in marriage negotiations through the 1990s had abated and exchange of sisters, though perhaps through 'more diverse paths of kinship connection' (2016: 166), was again being used to legitimise marriages. Many of these changes may lie in Suabi's future, should the promises of PNG LNG and other resource developments fail to materialise and the exploration that is feeding money and dreams into the community dissipate. But some changes Knauft described among Gebusi ring less true for Kubo. He wrote of gender relations becoming less fraught, with the wife-beating prevalent among Gebusi in the 1980s now less common. But Kubo men did not beat their wives in 1986–87. The only incidents of domestic violence we saw in 15 months were an occasion when a man struck his wife after he returned from an unsuccessful hunt and another, undoubtedly mutual, which resulted in scratches to the woman's finger and the man's nose. In seven months at Suabi, we knew of one case when a man returned drunk from Kiunga, and beat his wife for suspected adultery while he was away, leaving her with noticeable bruises; his behaviour was widely condemned. When Alice and Jeff fought, Jeff punched a banana stem, not his wife, and it was he, not her, who displayed (probably fictitious) signs of injury as a result. And while Knauft commented that 'shamanism, traditional spiritualism, and sorcery violence' had not re-emerged among Gebusi (2016: 135), at Suabi the former, expressed as curing dances, had re-emerged by 2014 and the latter two have remained though in somewhat modified form. Kubo no longer physically kill sorcerers—though we suspect one such killing did occur at a Febi community in about 2011, in mountains to the northwest—but they still hold sorcery divinations; several men we knew had undergone such tests in the past two or three years after being accused of particular deaths. Sorcerers, now, are killed by spirit-beings that mediums, often for commission, send out to avenge a death. In 2013, we attended the burial of a sorcerer, diagnosed as such because of the manner of his death. Another man, who we had known well at Gwaimasi, died in 2014 and was recognised post-mortem as the sorcerer responsible for the death of his own young son.

5. Navigating the Future

Zavia came to our house carrying several sheets of paper on which he had listed 150 names that he wanted us to type and print. This list, he said, would provide the basis for registering an Incorporated Land Group (ILG) as, under government regulations, this was necessary before benefits from the Liquefied Natural Gas (LNG) Project could be paid to those who proved to be eligible. But his list was puzzling. We expected that members of Zavia's clan would form the basis of the list, and we thought we knew most of these in person or by name. Our understanding was that the clan Yawuasoso comprised two geographically separated sections, one to the west of Suabi, close to the Strickland River, and another to the north of Suabi across Osio River in territory attributed to Febi-speakers. Zavia was affiliated with the latter. We knew of 39 men, women and children who were affiliated with this branch of Yawuasoso; the Strickland River branch included one man and his six children, and three or four additional children could have qualified as Yawuasoso by adoption. Even if all had been included, Zavia had listed three times as many people.

For each of the 150 people on his list, Zavia had recorded 'name', 'surname', 'sex' and 'age'. Under 'surname' we expected to see either father's name, husband's name or, because some people used a different convention, the name of the senior male in the household where the person resided. At best, and with some creative thinking, we could link 19 of the names on Zavia's list to people that we knew—and only four of these were Yawuasoso. A few were the wives, or former wives, of Yawuasoso men. The rest of the 19 were people from at least six other Kubo, Febi or Konai clans together with one person from Morehead in the southwest corner of Papua New Guinea (PNG).

'Who is this?' we asked, pointing to the first name on the list—Zavia Habufa. As we suspected, it was Zavia himself. We had thought his father was Yameka, and had not known that Yameka's brother Habufa died soon after Zavia was born and that Yameka had married Habufa's widow. Yameka was Zavia's stepfather but he, himself, was not listed. Indeed, the name Yameka appeared only once, as the attributed surname of the man from Morehead. The second name on the list was Zavia's wife, Mefus, but the third and fourth—a boy and a girl reported to be one year old—were not known to us. Their surname was Zavia. 'They are my children,' Zavia said. But Zavia had been married for less than a year. His wife was pregnant and would give birth in a month's time but, as yet, there were no children. Zavia explained. He did not know whether the expected child would be a boy or a girl and he did not want to leave that child off the list. So he had advanced the date of birth and, to be sure, included one child of each sex.

'And who are these?' We pointed to the last four names: Dimogai, Ebine, Diremi and Kofainsa, all with the surname Boboko. Zavia told us that Boboko was his stepfather Yameka, Dimogai and Ebine were his stepsisters Robyn and Livia, Diremi was Robyn's out-of-wedlock infant child Jeremy, and Kofainsa—'my small brother' Zavia explained—was the son of a deceased man from the Strickland River branch of Yawuasoso. The name Diremi was simply a variant on the spelling of Jeremy. Boboko, Dimogai, Ebine and Kofainsa were 'custom' or *tumbuna* names. Most people at Suabi seemed to have at least two names—one in the local language and another that was English—and preferred usage by others switched back and forth through time or varied with context. But we had not previously encountered the names Boboko, Dimogai or Kofainsa. They had made their appearance for the purpose of Zavia's list. Indeed, the last of these three names was associated with Yubi, a mythologically important, crocodile-infested lake on the land of the northern branch of Yawuasoso.

With Zavia's guidance we could now put faces to many of the other names on the list. We could identify 18 purported members of Yawuasoso clan—though that tally included the two named but unborn children. For these and many other people, Zavia simply had used names we did not know, and that did not have currency in conversation. Often, however, he combined a person's custom and English names, treating one as 'name', the other as 'surname'. Sometime he used the name of a person's grandfather as their 'surname'. And, sometimes, as with Kofainsa, he gave the name of a mythological being to someone who had never before been known by that name. But, in devising his list, his reach was further. Most of the people he included lived far from Yawuasoso land. For example, he included geographically distant kin of his mother. In the late 1960s, Kubo people living immediately west of the Strickland River hosted a party at Sigiafoihau (see Chapter 3). Their guests included Habiei people, five of whom were killed and eaten. But at least two young girls—they were sisters—were spared, and taken as wives by Kubo men. Zavia's mother, Gisio, was a child of one of those marriages. Zavia had listed a Habiei man—Gisio's classificatory brother—and four of that man's children.

There was additional information on Zavia's list. He had assigned 'land rights' to 21 of the named people. All who were Yawuasoso were in this set. Other people were classed as 'members' of the intended ILG; they could use the land in question but would always be beholden to those who held 'land rights'. Later, Zavia provided a cover page for his document. Here he had written the names of five men who would serve the ILG as 'Chairman', 'Vice-chairman', 'Secretary', 'Treasurer' and 'Committee'. Only two were Yawuasoso—Zavia was Chairman—and, Zavia aside, none of the others featured on his list of 150 names. Of greater significance, however, was the fact that Zavia's document made no reference to Yawuasoso. The cover page was unambiguous. The list was on behalf of a newly devised subclan (Tafenfen) within a recently designated major clan (Mogotie) within the purported 'tribe' Fembi (Febi). What had once been a named area, perhaps a named subclan, within the land of Yawuasoso was now upgraded to the status of 'major clan'. But there was no listed person who, by birthright, was affiliated with that area of land so, Zavia and others told us, that clan had been 'given' to him; it was now his task to oversee the allocation of subclans within Mogotie and membership within the subclan he had chosen for himself.[1] In a court of law, all the names on the list could have been judged invalid. And Zavia himself, though he had visited Yawuasoso land, had never lived there. From early childhood he had lived at Suabi. His first language was Kubo, not Febi.

Through the early months of 2014, at Suabi, people devoted much time and effort to preparing lists of names like the one Zavia brought us. It seemed that the entire landscape of social relationships was being redrawn. Two decades earlier, at Gwaimasi and Mome Hafi, *oobi* identity had been grounded in the interactions of particular people with others and with place. Now, it seemed, clan 'membership', as encoded in those lists, was something that could be bestowed on someone or withheld from them. And where, before, boundaries had been mutable they were now being fixed on paper. In seeking paths to a barely imaginable but greatly desired future of vast wealth and influence, people at Suabi were seeking to navigate the complexities of government and company bureaucracy— complexities that were ever-changing, impossible to pin down and could be known only by paying constant attention to rumours that trickled

---

1   One woman was said to be the only living person in the group named Mogotie. She was not included in Zavia's list but was named as the 'representative' of a different subclan within Mogotie.

in from outside. In this process, however, people were exploring very different ways of being in the world; indeed, they were bringing into being different kinds of people and different kinds of social entities. These changes, and the navigation of imagined futures that frame them, are the themes of the present chapter.

Constructing Lists

Drawing up lists of people was not new to people at Suabi. 'Class lists'—names of all pupils in each class—had been prepared each year by teachers at the school that had opened in the late 1980s. These provided a familiar model when people started to draw up their own lists. Indeed, on the earliest lists the names of people, both adults and children, were listed under the heading 'Pupil'. The first list we saw had been drawn up by members of a clan that held land adjacent to the Baiya River, downstream from Suabi towards its junction with the Strickland River. It was a 'family list', prepared at the request of International Timbers and Stevedoring (IT&S), which was seeking access to local land and, to this end, needed documentation of who was entitled to grant permission for that access.

In the mid-to-late-2000s representatives of IT&S visited communities east of the Strickland River. The understanding of many Kubo people was that this company planned to build a road—a 'Trans Papuan Highway'—that would commence at Kiunga in the west, pass through Nomad (25 km south of Suabi) and continue to Port Moresby. The distance to be traversed, across large rivers, through swamps and forest, would be of the order of 600 km. It was understood further that side roads would link Nomad with Suabi, continue north to Juha and turn west to Komagato on the Strickland River. The chief executive of the company—Neville Harsley was known to Kubo people—stated that IT&S would develop cash crops, timber plantations and a sawmill in exchange for rights to selectively log up to 5 km either side of the road (Chandler 2011a, 2011b). He had offered at least one Kubo man an all-expenses paid trip to Port Moresby where the lease would be signed.

IT&S's highway proposal fell within a Papua New Guinean 'lease-leaseback scheme' under which customary land rights may be converted into legal titles, then transferred to the National Government to be transferred to private companies as Special Agricultural and Business Leases (SABLs). For the duration of the lease 'all customary rights in the land, except those which are specifically reserved in the lease, are suspended' (Filer 2011a: 2).

A Forest Clearance Authority covering the Nomad area was issued to IT&S in November 2010 (Mirou 2013: 120). The lease would stand for 99 years and was modified to include vastly more land than that entailed by a 10 km-wide corridor along the proposed road. It now embraced virtually all the land of several Western Province language groups.

Across PNG there has been a dramatic increase in take-up of SABLs since the year 2000, and considerable concern that proposals for local 'development' within, or near, the areas covered by the leases have not come to fruition (Filer 2011b; Nelson et al. 2014). In 2010 the National Government instituted a Commission of Inquiry into these leases. Reports from the Inquiry were submitted to Parliament in June 2013 (Mirou 2013; Numapo 2013). Of 42 leases examined in detail there were only four cases in which the consent of genuine landowners had been obtained and in which a viable agricultural project was initiated. With specific reference to IT&S the Commission found that the company had been 'directly manipulating the SABL lease back process' and had done so in ways that were 'encouraged' by the 'gross negligence' of agencies of the State (Mirou 2013: 402).

At Suabi, people did not understand the intricacies of the SABL system. In 2011, most were excited by the prospect of a road. They resented the fact that two Europeans—the long-term missionary Tom Hoey, and a Kiunga business man and former government minister, Warren Dutton—were publicly opposed to the lease, arguing that those men wanted to keep people as they were, backward and without 'development'. The list we were shown had been drawn up with the intention of complying with the wishes of IT&S. It was prepared on behalf of Mithy clan and, under the heading 'Pupil', named 60 people, most of whom were members of Mithy, Dabamisi or Buwo clans. The land of Dabamisi clan is contiguous with that of Mithy and a recent marriage between a Mithy male and a Buwo female rationalised inclusion of members of the latter clan. Wives to members of these three clans were also included, though they were assigned 'user rights' rather than 'full rights'. Two deceased people were listed as heading one household, and a woman from Ialibu, in the Highlands, was listed as the second wife to one Mithy male. This marriage had not in fact occurred, nor had the couple ever met; their 'love affair' was via mobile telephone and the intention was that they would marry when the man concerned had become wealthy and could pay the required bridewealth. He had expected that the IT&S project would provide him with immediate financial benefits.

The second list that we saw was motivated by a claim to land that had already been alienated rather than by a claim that would enable alienation. A man named Michael asserted that he and his kin were the true landowners of the area at Suabi on which the airstrip, mission station, school and community health centre have been established. His listing of 86 names was one of a number of actions that he took, over several years, to promote his claim. His target audience was, variously, other Kubo people and visiting outsiders. In 2013, near the airstrip and facing the exploration camp, he erected a large sign that comprised two paintings— one a rock with an inscribed face, the other a landscape that located Suabi relative to mountains and river (Fig. 5.1)—and the following words:

Welcome to Soabi Airstrip, Baiya River

Tibidibo is my name. I am a sacred stone with a human face. I come from my sacred swamp called Sodu'u. My home is in the Baiya River adjacent to the Soabi Airstrip but when the Missionaries came to Soabi and built the airstrip, they destroyed my sacred home. This is my home and inheritance and nobody takes my right away from me. AS LONG AS I EXIST, I AM TIBIDIBO.

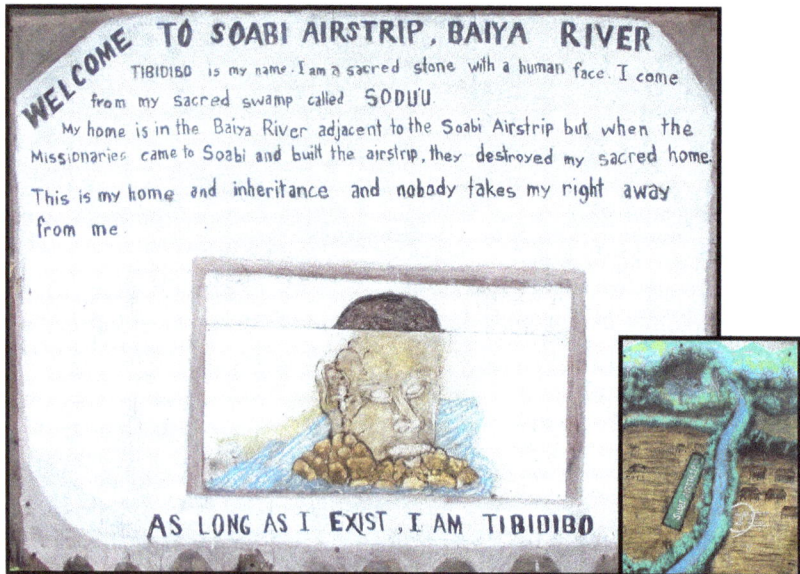

Figure 5.1: 'Tibidibo is my name.' Sign erected at Suabi airport.
Source: Photograph by Peter D. Dwyer, 2013.

Few people at Suabi paid much heed to Michael's proclamation. One commented cynically that 'Company will go. The trees will grow back. The sign will be covered. No one will know it was there. It doesn't do anything'. And his assertion of 'ownership' over the airstrip land was similarly ignored; several people told us it was simply wrong, though none overtly challenged him about it. Certainly, exploration companies had little interest in Michael's clan list and associated map. Money paid for the use of land beside the airstrip—the base camp was close to what all acknowledged had been Tibidibo's home—was handed to representatives of three clans, not just to Michael's line.

In 2013 and 2014, however, the focus on drawing up lists of names increased dramatically.[2] The stimulus now was the PNG LNG Project and the requirement, as local people understood it, that financial benefits would accrue only to those who had registered an ILG. But the people had little reliable information about the registration process, and had not seen the necessary registration forms. They knew that a list of prospective members was to be prepared, and that a management committee and a dispute resolution authority had to be nominated. But there was much less certainty about the form that any of these should take. For this, they turned to rumour. They gathered that prospective members were to be identified by name, age and relationships to others, and could be differentiated according to rights held, but were unsure how these were to be recorded. They were, however, sure that each list should comprise exactly 150 names, and that no name should appear on more than one list.

With these minimal guidelines in mind, the task of drawing up lists began (Minnegal et al. 2015). In many cases, a single person—the sole member of a newly designated 'subclan'—was responsible. In other cases, a senior man advised a literate son, daughter or wife who to include. Sometimes members of a major clan met, to compare notes and decide who should be included in which subclan and under what name. But the eventual lists were drawn up by individuals, young men for the most part, who could be seen crouched over paper under a house, head in hand, sighing loudly as they agonised over the task. Both Febi and Kubo people drew up lists. (It seemed that earlier ILG applications by Febi people had not come to fruition.) But people from neighbouring language groups—Konai to the west, Samo to the south and Bedamuni to the east—also prepared lists.

---

2   In the first four months of 2014, we typed 30 different lists, saw another 20 complete or partial lists, and typed an additional seven lists detailing the Executive for intended Incorporated Land Groups.

In all cases, people strove to include the names of 150 people; though not specified in the Land Group Incorporation Act, people insisted that they had been told this was the number needed.[3] We ourselves suggested that it was not necessary, that elsewhere in PNG people were registering ILGs with relatively few members (Weiner 2007; Bell 2009; Goldman 2009: 3.13). The unanswerable response was that our own understanding was prejudiced by Australian law while they were striving to comply with PNG law. We conceded ignorance and typed and printed the lists as we received them.

Though the layout and terms used differed somewhat between lists, they were all presented as an assemblage of families, each comprising a married man, identified as 'head of family', his wife (or wives) and unmarried children. Married sons were listed as heads of their own family, while married daughters were usually excluded on the expectation that they would be listed elsewhere as part of their husband's family. The importance of 'family' to status was immense. The overall structure of the lists conveys a strong impression of patrilineality, at odds with the distinctly cognatic orientation of much past and present practice. This was reinforced by comments from some people that 'under PNG law, only men can own land'. In almost all cases, the 'father's name'/'surname' listed for wives was that of their husband, not of their father. But closer analysis reveals a more complex picture. The lists included not just agnates—families of men clearly identified with the subclan, or those of 'brothers' from the same clan—but also a range of affines—the families of mothers or even wives' brothers. In two cases, in-married men were included: former community health workers from Morehead and Oksapmin who had married local women while posted at Suabi and still spent much time in the community. In several other cases, lists included natal families of in-married women from other language groups.

---

3   We suspect that someone, taken to be knowledgeable, had said that it was inadvisable to include more than 150 names on an ILG list and that this was misheard, or misunderstood, and taken to mean that each list should include that many names. Jorgensen (1993: 69) reported that, at Tabubil, in the 1970s and 1980s the local government council, encouraged by government employees, passed a set of rules that aimed to limit bridewealth inflation. Among Telefolmin villagers, however, these rules were interpreted as setting a minimum rather than a maximum bridewealth payment.

As with cases mentioned above, some lists also recorded marriages that had not occurred and attributed children to people who were not their parents. There was no consistency in the attribution of 'full rights' or 'user rights' to named people; some lists accorded full rights to everyone, others to all males but not to females, while other lists, like that of Zavia, were more selective. For people older than about ten years, attributed ages were a matter of guesswork and often wildly inaccurate. (For example, the documented ages of many children and their mothers were such that births had purportedly occurred when the woman was herself an infant.)[4] Attributed birth order, however, usually matched what we knew from our own earlier records and was clearly more salient than any concern with actual year of birth. Though no name appeared more than once, in many cases the same individual was recorded on two or more lists under different names. Two brothers, who produced separate lists, each named their father and mother together with six of their parents' children but, in all cases, the attributed given names and surnames differed between lists. One list was accompanied by a brief origin story, four with logos that represented clan stories or clan resources, one with a map depicting the border of clan land (Dwyer and Minnegal 2014: 42, Fig. 2) and one with a detailed account of waterways, ridges and gullies that marked the border together with lists of 60 mountains, 83 creeks and sago swamps and 83 rivers and creeks found on clan land. These last two lists were the only ones we saw that specified the land that was taken to be associated with the group in question.

Few of the listed names of people were fictional. In the struggle to find 150 names—and several people spoke of how difficult this task was—their reach was broad. Despite the use of multiple names for the same person, and the extension of lists to include allies and affines, there were simply not enough local people to fill lists of the desired length. So they were expanded in other ways. People who had attended school at Kiunga or in the Highlands listed members of families with whom they had boarded,

---

4   There is no formal system at Suabi for registering births and deaths. At best, and very intermittently, resident community health workers have kept records—125 recorded births and 26 recorded deaths since 1999, with no birth records kept in the years 2001–02 and 2005–07—but there is no system whereby these are officially lodged and record books, if maintained at all, are often lost. There is now, however, much interest in date of birth among young men and women. Some were very excited when we could tell them the date, based on our earlier field records, and others were disappointed when we could not. Some people assume that it not merely the date but also the day—a Monday or a Friday for example—that is carried forward from the year of birth into the future.

or who had otherwise cared for them, during those years. People who had travelled to Tari, Mount Hagen, Goroka, Lae or Port Moresby listed people who had facilitated their journeys by feeding them or providing accommodation. And one man listed the families of people he had come to know through his long experience with the Seventh-day Adventist Church—his list extended to Manus, an island 700 km northeast of Kubo land. The outcome was that thousands of people were listed as having an affiliation with land that they had never visited. The extreme case was probably that of Simiti. He showed us the lists he had paid someone in Goroka to draw up under his guidance. There were ten of these, each representing a recently devised subclan within a recently devised major clan of Febi, and each included 150 names. None of those names were duplicates though many were alternative names for the same person. Simiti's lists recorded 1,500 different names as being associated with the land that he claimed as his own. His immediate kin—those in the assemblage Wuo which had recently fragmented as eight major clans—were engaged in the same exercise. The lists produced by these groups alone were likely to comprise 9,000 names but the entire Febi population is less than 500 people.

We asked several people how they intended to distribute money once they received it. Would they share to all the far-flung people they had named? One said that he might 'throw them a few coins'. The others responded that they would give them nothing, that those people did not know they had been listed. We ourselves were asked, more than once, whether we would like to be listed. We declined and, again, appealed to legality. 'But no one would know,' one man explained. 'The list is secret.' There were, however, people from distant places who did know they had been listed, and some who grasped the opportunity this might afford. One woman from the Highlands applied to ExxonMobil for a grant to support her enrolment at the University of Queensland, attaching a letter of acceptance from the university stating that her enrolment was contingent upon fees being paid in advance. She declared in her application that she was a landowner in the Juha area—naming the clan of which she was a 'member'—and thus entitled to benefits from Juha. But, in general, people were cautious. When naming people who would comprise the Executive of their intended ILG they listed only local people, often immediate members of their families and often, because again the number of people available was few, they 'appointed' children, less than ten years old, to positions such as 'secretary', 'treasurer' or even 'Chairman'.

At Suabi, it seems, the process of drawing up ILG lists was informed by deeper understandings of personhood. Persons here are made, and remade, throughout their lives as they interact with other beings, other persons around them. And thus, to those engaged in their preparation, the lists were a right and proper record of all who had a stake in the world they inhabited. But clearly, too, it is recognised that not all contributions are equal. Indeed, in the ways the lists were conceived and constructed, categorical distinctions—between heads of households and members; between men and women; between married and unmarried persons; between 'owners', 'users' and those who have no place on the list—have been reified in ways that were perhaps not intended but will undoubtedly have ongoing effects.

Reconfiguring Groups

While the lists that people produced were problematic with respect to the names they recorded, there were implications of greater importance at a broader scale. These entailed realignments and reframing of social groups in ways that were very different from what had been in place before. These changes, and their present and future consequences, may be traced to two primary causes: PNG bureaucracy and one remarkable man, Bob Resa. We should start with Bob, whose actions precipitated a dramatic series of realignments.

Division and Amalgamation

In 1995 we walked west to east through much of Febi territory. Two weeks into our trip we spent one night at Siabi, the village nearest to the Juha well heads. Through the previous days the Siabi community had hosted a *tok save* school at which people were taught Tok Pisin. The event finished with a feast. But a dispute had arisen. Bob, who had for several years guided the community in the ways of the Christian Brethren Church (CBC)—he himself had spent time at a CBC mission in the Sepik—was a central protagonist and, in keeping with common practice, responded to the dispute by departure. With his wife, first-born son and others he headed south to Suabi. We caught up with him on the track. He carried his young son on his shoulders and was teaching him Tok Pisin: the language, Bob told us, that Jetha would need in the future. For the next 12 years Suabi was his primary place of residence. He ran a small trade store, wondered about the potential for tourism, established a small plot of agarwood,[5] and obtained funds with which he bought the equipment—a rice mill and a portable saw mill—needed to initiate small-scale rice farming and walk-about logging. There was, however, no external market for rice and sawn timber, and these two ventures failed. Bob's wife died, he remarried and fathered three daughters. By the mid-2000s he was engrossed in the potential of the PNG LNG Project. He was, after all, of Wuo clan—the least ambiguous of landowners in the Juha area.

In 2005, under the guidance of Huli patrons, Bob registered a landowner company with 11 named directors each representing a purported Febi clan. Through the next two to three years Juha Development Corporation (JDC) provided casual labour associated with drilling of two of the Juha wells (Goldman 2009: 3–101; Jackson 2015: 66). The work was managed by the Huli company Gigira Development Corporation (GDC). In late 2007 or early 2008, JDC received PGK810,693.13. The money was distributed to 11 named men; three (including Bob) had been listed as directors of JDC, three were other Febi men and five were Huli men at least some of whom had connections with the directorship of GDC, but eight of the original named directors of JDC received no money. In May 2008, JDC was deregistered and, soon after this time, encouraged by Huli patrons, Bob left his wife, married a Highlands woman, declared himself 'Chief of the Febi Tribe' and relocated to Port Moresby.

By October 2008, with several other Febi men, Bob had registered another company and, thereafter, from time to time, sometimes with others, sometimes alone, he registered yet other companies. He told people that, in Port Moresby, he could keep a watching brief on bureaucratic and legal problems, and would be better placed to ensure that Febi and Kubo people would eventually receive what was rightly theirs. Through the next five years, Bob's companies accessed more than two million kina as Infrastructure or Business Development Grants. This money was diverted to living in Port Moresby, purchasing a house, leasing office space, supporting a coterie of young male acolytes who assisted him

---

5   Agarwood is a species of *Aquilaria* (also known as eagle wood, gaharu, 'gold tree' or the 'wood of the Gods') that produces a dark resinous wood in response to fungal infection. It the Middle East and Asia it is used as incense and for medicinal purposes and has a role in various religious ceremonies. The resinous wood may fetch up to USD30,000 per kilogram. With a decline in availability in Asian countries, there has been increasing attention to sourcing the wood from PNG. In 2001, payments to Papua New Guinean harvesters varied from PGK371.66 to PGK1,139.5 per kilogram, depending on grade (Zich and Compton 2002; Gerber and Roberts 2005).

with preparing documents and translation (he could not write or speak English), hiring legal advice and ensuring that his daughters continued their high school education. None of the money was directed to developments within the territories of Kubo or Febi people and, indeed, local people who had paid to become shareholders in Bob's companies did not receive dividends. More seriously, however, rumours began to filter through that Bob had 'given' his Huli patrons rights to two of the five well heads at Juha.

In December 2012, Bob returned to Suabi. His kin at Gweobi Corner had organised a 'party' at which numerous pigs would be exchanged, and used this to summons back from distant towns not only Bob but also others who were living elsewhere. People at Owabi Corner, concerned that those at Gweobi would build alliances that excluded them, decided to hold a series of exchanges at the same time, and people at Suabi Corner followed suit. All summonsed their kin to return and participate. Most responded to the call, some walking five days from Kiunga, others walking from the Highlands. A few, like Bob, had not been back to Suabi for several years, attending high school in Goroka or technical college in Lae, or simply caught up in the excitement of town life. At Suabi, anticipation had been building for weeks; more than 65 pigs had been rounded up for the occasion, to be killed, butchered, cooked and the meat exchanged.

Bob had chartered a plane for his return and, as he stepped from it, filmed the many people waiting to greet him. He had insisted that the feast should be accompanied by a 'cultural revival'—mock raids, costumed dancing, hilarious performances by men disguised as ogres, and speeches in which he himself would talk of the future. He filmed these activities, telling us that he would show the outside world that the people of Juha were 'primitive' and 'remote' and needed support. From a purpose-built stand overlooking the festivities, he spoke of all he had done, and was doing, for the community, hinted at planned autonomy for the Juha area and the desirability of electing a 'Juha president'. His audience was cynical. There was no direct confrontation but the mood was apparent and Bob felt it; he spoke to us of his distress at sensing that his efforts were not appreciated. But others spoke to us too, of their sense that Bob had failed them. He had been their 'pipeline' to the outside world, to the bureaucratic procedures that they needed to navigate but did not understand. He represented them, but there was nothing to show for it. He lived the 'good life' in Port Moresby, with access to money, with a succession of wives—he had married a second Highland woman—and travel to Australia, while they remained a 'remote' backwater. 'Development' did not come. Bob had been a 'good man', but now he was a 'big shot'.[6] He even looked the part. People observed that he was fat, that he neither ate nor worked as they did. In dress and mannerism he was modelling himself on PNG's first Prime Minister, Michael Somare. But fatness was a recognised characteristic of *sanguma* (magic-men), and there were growing murmurs within the community at Suabi that those who—like Bob—chose to spend their time in towns seemed at risk of becoming *sanguma*; they were, in a sense, 'supping with the devil', and the effect could be seen.

Bob's brief return to Suabi in late 2012 initiated a cascade of changes. In 2014, when we conducted a population census, other Febi-speakers insisted that he should not be counted as a Suabi resident. If he came again, people said, he would be charged rent or could sleep beneath coconut palms that he had planted years before. Members of his own clan—Wuo—remained appreciative of Bob's earlier contributions, but concluded that he was now too powerful, that he no longer had their interests at heart. They could not remove him from his position as a senior member of Wuo, so instead they decided to marginalise him. And they did so by asserting that eight subclans within Wuo should each be recognised as a distinct 'major clan'. Bob would remain 'Chief' of one of the eight, be acknowledged as owner of the land around one of the Juha well heads, but he would have no rights to, or right to speak for—much less give away— the other well heads. These were on the lands of other 'major clans', for each of which another man was appointed as 'Chief'. This was done in advance of, and in preparation for, an expected clan vetting process that would be organised by the Department of Petroleum and Energy (DPE) in November 2013. But the decision to restructure Wuo had implications for other groups. It was thought that benefits from the PNG LNG Project would be distributed as equal shares to either recognised major clans or recognised subclans within major clans. No one was sure, but it was necessary to be ready. Headubi reorganised as four major clans, Yawuasoso as three, Gumitie, Kesomo and Mora each as two. And within these five assemblages—all major clans, people had been told, should have

---

6   Keir Martin (2013: 3, 123) writes that Tolai use 'big shot'—*biksot*—as a derogatory label for men who they consider 'have forgotten their obligations to others in their eagerness to join the ranks of an emergent socio-economic elite' and whose engagement in ritual may be commercially motivated. People at Suabi did not use the expression but the sentiments they expressed were in keeping with it.

subdivisions—there was now a total of 131 named subclans. None of the latter had existed as named social groups before. Given the size of the Febi population in 2014, the average number of people in each of these newly devised subclans would not reach five. As Ernst (1999) reported for Onabasulu, people were deeply engaged in a process of 'entification'. The 'entities' they created were eventually accepted as legitimate by DPE.

Many of the divisions that produced these new entities were made at very short notice, and with minimal consultation. Zavia, then a young man of 25, was the only representative of Yawuasoso at the clan vetting forum. He saw what others were doing, and decided to split his clan. He told us of his fear and uncertainty about whether he was doing the right thing, of coming back from the meeting a few days later and having to tell his older clan brothers of what he had done. He was greatly relieved when they approved, telling him he had done well. But all must now live with his decisions—decisions not only about splitting his clan but also nominating subclans—each associated with a focal site—for each of the new major clans, and then determining who would belong in which of the new segments.

For those who were creating subclans there needed to be either some sense of prior existence as a named group of associated people or some sense of past association with a particular area of land where the people concerned, and their forebears, had often planted gardens, harvested sago or hunted. But the potential members of these subclans were few. At the same time there were people in other groups, particularly non-Febi groups, with a different problem; they might not be recognised as having rights to benefit from developments on Juha land. The solution arrived at was to the advantage of everyone. Clans with little claim to the area where gas wells had been drilled, or where the proposed pipeline would run, were invited to affiliate with clans that were understood to hold unambiguous rights to those places. In the ILG lists that these groups drew up, they declared their 'clan'—it might be Kubo, Samo, Konai or Bedamuni— to be a subclan within a Febi or a Kubo major clan of Juha beneficiaries. In the former case, they usually asserted that they were members of what they named the 'Febi Tribe'. Primary realignments that followed from these decisions are summarised in Figure 5.2.

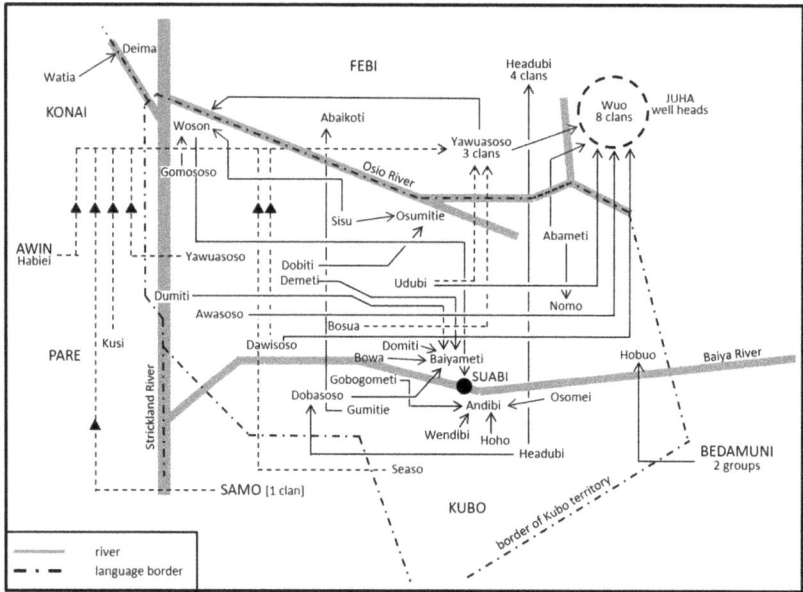

Figure 5.2: A schematic representation of social realignments.

Source: Authors.

Note: The figure shows realignments of Kubo clans or subclans, together with some Awin, Pare, Samo and Bedamuni groups, within Febi clans or other Kubo clans as revealed in the process of compiling lists of members of ILGs. To enhance clarity, arrowed lines converging on Yawuasoso are dashed.

Most of the realignments shown in Figure 5.2 are with the eight clans now representing Wuo and the three clans now representing Yawuasoso. Thus, Awasoso and parts of both Dawisoso and Udubi—all are Kubo—identified as subclans within the greater Wuo assemblage. Similarly, Seaso, Strickland River Yawuasoso, Bosua, parts of both Dawisoso and Udubi, and a Samo group all identified as subclans within the greater Febi Yawuasoso assemblage. Thus, while Wuo provided places for groups from some Kubo clans, in seeking to find members for a greatly expanded list of subclans Yawuasoso came to encompass people from language groups even further afield (Pare, Awin and Samo together with Kubo).[7] One member of a Yawuasoso group was named the official representative of the Kubo clan Woson, and another named as the representative of a subclan within the greater Wuo assemblage. (This last arrangement arose because the man in question had caused difficulties for his immediate family and his father was unwilling to assign him his own subclan within Yawuasoso.)

---

7   Ironically, in these cases the Kubo groups were seeking alignment with a named Febi group that, on the basis of recent subdivision, was no longer extant.

One Konai clan (Watia) realigned as a subclan of the recognised Febi group Deima, and finally, Abameti was accorded subclan status within the Kubo clan Nomo, though the senior Abameti claims affiliation with Wuo. One of that man's daughters has, in fact, been named as the representative of one of six subclans within a Wuo clan that is overseen by her husband.

What is evident in Figure 5.2 is that while Febi clans have tended to split—to create more clans—Kubo clans have tended to amalgamate (Minnegal et al. 2015: 506–7). The former is unsurprising given local understanding that there was little ambiguity about the status of Febi clans as Juha landowners; members of those clans could thus focus on strategies to maximise their own share of anticipated benefits. There is, however, more ambiguity among local people about the status of different Kubo clans as legitimate beneficiaries. Those who seemed to have at least some claim to benefits sought to reinforce their position by recruiting others to their cause, while those who feared missing out entirely sought to ensure at least some entitlement by aligning with groups whose status seemed more certain. Baiyameti and Andibi, for example, are Kubo clans with land that encompasses the airstrip and its approaches at Suabi. These clans have been recognised by Febi landowners as legitimate beneficiaries on the basis that, for three decades, the airstrip has provided the entry point and servicing base for the companies working at Juha. Baiyameti has only one living member, a woman. In an earlier generation, a geographically separate branch of this clan occupied land to the west, near the Strickland River, but this branch has no living members. The nearby clan, Dumiti, now 'speaks for' Baiyameti and, with reference to the PNG LNG Project, is taken to be synonymous with it. The referents of Baiyameti have been expanded further by according Bowa, Demeti and Domiti subclan status and, similarly, the referents of Andibi have been expanded by according Osomei, Gobogometi, Wendibi and Hoho subclan status. Demeti, however, is also regarded as a subclan of a greater assemblage named Dobiti which itself has, for ILG purposes, been included as a single subclan of Osumitie. Finally, the clan Woson—with land located in foothills near the confluence of the Osio and Strickland Rivers—has been included, along with Baiyameti and Andibi at Suabi, within the set of three Kubo major clans that are to share two per cent of benefits from Juha. Woson was included in this set on the basis that the senior male was a key translator during the period that benefit-sharing meetings were held at Suabi and Moro. Among others, however, Woson has granted subclan status to Gomososo (though not by that name) and Sisu, though the latter is simultaneously regarded as a subclan of Osumitie.

One other example warrants mention. Hobuo is a Kubo clan. As a child, in the 1960s, the now most senior Hobuo male was kidnapped by, and reared as, Bedamuni. He married an Etoro-speaking woman and commenced a family. In the early 1990s, with his family, he was living at Suabi and was being encouraged to reassume a Kubo identity. His oldest son attended high school at Kiunga and was killed there. In the intervening years, Sosogoli, with his family, moved back and forth between Bedamuni and Kubo land. In the 2000s, he established a community on his natal clan land, about 3 km northeast of Suabi. Most members of this community are Bedamuni. The Hobuo submission to DPE identified seven subclans, two of which were named as Mougulu and Adumari. There was potential ambiguity here, for Mougulu and Adumari qualify as both named Bedamuni clans and named Bedamuni villages. The villages include residents of many different clans, and the 2011 National Census recorded their combined population as 839 people.

Not everyone at Suabi was comfortable with what was happening. We asked Doiyo, the senior Nomo man, what subclan he belonged to. Nomo people were holding many long meetings at this time, and drawing up documents in which they named eight subclans. We had known of two of these for 20 years, but not heard of the others. Doiyo was puzzled by our question. He was 'Nomo'. He was unwilling to refine that classification. On a separate occasion we asked his oldest son Asnah the same question. He too said he was Nomo. We challenged him. We had, after all, seen and typed a list in which Doiyo and Asnah were named as the representatives of specific, different subclans. '*Nomo stret*', was Asnah's response; he did not identify with a particular subclan. To these men, identification with particular subclans was appropriate in the context of drawing up ILG lists but had not, as yet, been generalised to all contexts.

Samwan—a Morehead man who had married into the community—did not approve of attempts to provide 150 names on the ILG lists that people were compiling. People should give gifts to those who had supported them in the past, he said, but they should not give a permanent share of their income. Another young married man, Yasa, acknowledged that he was confused about the requirement for 150 names. He asked us to type two different lists, one with 150 names and the other more focused on kin with only 57 names. He would prepare both and, later, would listen to what government people said and know which list to provide. A third man, Mark, very recently married, was sure that people could compile lists with less than, but not more than, 150 names. But, to be safe, he would include the higher number on his own list.

Pastor Martin was exceptionally thoughtful and, at least to us, forceful in expressing his opinions. He was scathing about people declaring subclans within new major clans. 'How are they finding these? They are making things up. They are not following custom law.' He insisted that Wuo was a single clan, not a confederacy of eight major clans. 'When we see someone from there, we say Wuo people are coming, not any smaller name.' He argued that people should be cautious in drawing up their lists. People thought that vast amounts of money would be coming, enough for everyone, he said, and so they have focused on inclusion rather than exclusion. But no one knows yet how much money will come, it may be limited. He himself knows that among Pare-speaking people west of the Strickland River there is a 'fire clan'. They will be distantly—mythologically—related to him, but he will not be including them in his ILG list, or assigning them subclan status within Osumitie.

Opinions of this sort were in the minority and rarely voiced in public. But throughout our time at Suabi there was other evidence of discomfort. When our conversations were explicitly concerned with Juha and ILGs then the language of 'major clans' and 'subclans' was to the fore; the new divisions and amalgamations of social groups were central to people's remarks. But in less formal contexts—when we were able to put Juha and ILGs to one side, and that was not always easy—the language, and the social groupings spoken of, often harkened back to our earlier experiences and understandings. Sometimes, in the 'new' context, it seemed that the Kubo word *oobi* had been forgotten or abandoned, replaced by the Tok Pisin *klen* or the English 'clan'. In lighter conversations, however, when anxieties about accessing money could be put aside, the word *oobi* often resurfaced, and did so spontaneously. In these contexts too, it was more likely that people—women especially—would use people's *tumbuna* names, even in cases where the person concerned preferred that that name be forgotten.

Inclusion and Exclusion

In PNG, petroleum prospecting, retention and development licences are granted over areas of land that comprise a set of 'graticular blocks'. These blocks are predetermined as areas delimited by one degree of latitude and one degree of longitude, and each has a unique numeric identifier. In drawing up ILG lists many people identified the block with which they asserted affiliation. Irrespective of the actual location of their own clan land, nearly everyone who identified a block nominated either the block on which the

Juha well heads are located or a block to the immediate southeast that is traversed by the proposed pipeline. Further, at the clan vetting meeting in 2013, people who had not affiliated with Baiyameti or Andibi asserted that Siabi, where the meeting was held and the nearest village to the well heads, was their place of residence. People were anxious. They were doing all that they could to establish their credentials as legitimate landowners. In those endeavours, to the extent that they understood what was expected they complied with the bureaucratic requirements of the State. The lists of names, clearly delimiting membership of named groups, and the aligning of those groups with government-designated classification of lands and locations, were undertaken in an effort to render themselves visible to the State, and to the companies that held control of hoped-for benefits. But the logic of government was not theirs. Nor were the bureaucratic requirements clearly understood.

It is clear from the foregoing that many individuals have been named on more than one intended ILG list. Similarly, many clans, either in their entirety or in part, have been subsumed as subclans within more than one of the current catalogue of major clans. What, however, was the basis for inclusion or exclusion? The reasons people offered varied. Affinal connections, long-term friendships, indebtedness arising from support in the past, or anxiety about past offences and the concern that there would be 'trouble' if a particular person or group was not included—these all influenced the likelihood that a person or group of people would be listed. Equally, current or long-standing tensions, or unresolved disputes, influenced the likelihood of exclusion. Some people were invited to be listed. Many did not know they were listed. And some asked to be listed. When two Mithy men, for example, asked to be listed within a subset of Wuo the man they approached responded dismissively 'who do you associate with, Koli clan or who?' He was reminding these Kubo men that they had few close connections with any Febi group. Though he added, more generously: 'Sorry, my lists are finished'.

We have seen that Wuo men sought to marginalise Bob Resa. After the clan vetting meeting, they sought also to marginalise most other claimants. One man drew up a document in which he listed, separately, the 'Principal Land Owners' and the 'Land Users' of areas encompassing the Juha well heads. The first list included 11 major clans within the Wuo, Headubi and Koli assemblage and specified the well head with which each one was associated. The second list, including the names of 27 major clans, was, in effect, an assertion that these groups may have used land in the vicinity of the well heads but they did not do so by right. In drawing up these lists, this man was marginalising people and clans with whom, to this time, he had had close personal and political relationships. Increasingly, people were negotiating in secret. Trust was breaking down. The effort to ensure access to future money was disrupting internal relations among Kubo and Febi people.

In the years that followed the 2009 Moro meeting (see Chapter 3), Kubo and Febi people, working within the constraints of an unfamiliar bureaucracy and a more familiar social world, reordered the latter as a means of formally representing themselves to outside authorities and ensuring their legitimacy (compare Jorgensen 1997). They were anxious, they were strategic and they often produced representations that ran counter to previous social arrangements. In part, they were successful. The *Recognised Major Clan and Sub-clan List* released by DPE in April 2014 conformed, with few exceptions, to their presentations at the November 2013 clan vetting meeting.

But their success was limited. A detailed comparison of the clans listed as entitled to benefits, and the share each could expect, in the 2009 Moro agreement and the 2014 DPE document is provided in Table 5.1. There are many differences. The number of recognised major clans has increased from 24 to 53 and the benefits expected to flow to Febi and Kubo clans has decreased from 92 per cent to approximately 65.5 per cent of the total. Insistent lobbying in Port Moresby by Hela Province claimants had paid off. Two Kubo clans that were not named in the Moro agreement appear in the later document (Andibi and Baiyameti). Himiti, which was named as a Kubo clan in the Moro agreement (though it was not known to us as a stand-alone *oobi*), has been reclassed as a subclan of Iodibi, with the latter itself now classed as a Region 1 clan—that is, as having very strong rights in the area of the Juha well heads. Strong affinal connections with Febi and an experienced Iodibi lobbyist explain this listing. A few names were missing from the DPE document. Osumitie, for example, was not listed as a subclan within the major clan of the same name. DPE officers accepted Martin's listing of subclans, but eliminated Osumitie because there were no living members. The Kubo clan Udubi was not recognised in its own right, though people in that clan were accommodated elsewhere. And Sisisti, a long-standing subclan of Nomo, was missing from the list. Some people suggested this was deliberate, asserting that DPE officers had eliminated this subclan because the man who represented it had been too outspoken. He had irritated them.

Table 5.1: Proposed distribution of benefits from Juha gas wells.

Moro agreement December 2009		Language	Recognised Major Clan and Sub-clan List April 2014		
CDOA equity [%][a]			Major Clan (subclans) [region] Benefits by clan/subclan [%]		
Febi clans					
Uwo	15	Febi	Wuo Fofasoso (10)	[1]	2 / 2.25
			Wuo Masumitie (10)	[1]	2 / 2.25
			Wuo Kulimitie (5)	[1]	2 / 1.13
			Wuo Kokora (7)	[1]	2 / 1.58
			Wuo Tigidin (5)	[1]	2 / 1.13
			Wuo Masisomitie (10)	[1]	2 / 2.25
			Wuo Manatie (6)	[1]	2 / 1.35
			Wuo Siyogutie (7)	[1]	2 / 1.58
Hiadibi	15	Febi	Hiadibi (4)	[1]	2 / 0.90
			Huisi (4)	[1]	2 / 0.90
			Barami (7)	[1]	2 / 1.58
			Mendo (7)	[1]	2 / 1.58
Koli	15	Febi	Koli (8)	[3]	2.71 / 2.98
Yawasoso	5	Febi	Dobametie (7)	[1]	2 / 1.58
			Yawososo (7)	[1]	2 / 1.58
			Mogotie (7)	[1]	2 / 1.58
Gumitie	5	Febi	Gumitie (5)	[1]	2 / 1.13
			Abain Kotie (7)	[1]	2 / 1.58
Howotie	5	Febi	Hwotie (7)	[1]	2 / 1.58
Kesomo	5	Febi/Kubo	Kesomo (9)	[1]	2 / 2.03
		Febi	Tebesutie (7)	[1]	2 / 1.58
Diema	5	Febi/Konai	Deima (7)	[1]	2 / 1.58
Botie	5	Febi	Botie (7)	[1]	2 / 1.58
Ulatie	5	Febi	Ulatie (7)	[1]	2 / 1.58
Osumitie	5	Kubo	Osomitie (5)	[1]	2 / 1.13
Mora	5	Febi/Tsiani[b]	Mora (8)	[3]	2.71 / 2.98
		Febi/Huli[b]	Amodo (7)	[3]	2.71 / 2.61
Not listed in Moro agreement		Febi	Bilatie (8)	[1]	2 / 1.80
Other Western Province groups–Subclan (Tribe)					
Nomo	2 [0.33 each]	Kubo	Nomo (7)	[3]	2.71 / 2.61
Hubowo		Kubo	Houbua (7)	[3]	2.71 / 2.61
Himiti		Kubo	Now included as subclan within Iyodibi		
Yiodibi		Kubo	Iyodibi (4)	[1]	2 / 0.90
Udubi		Kubo	Not listed as recognised clan		
Woson		Kubo	Woshon (6)	[4]	0.67 / 0.71
Not listed in Moro agreement		Kubo	Andibi (5)	[4]	0.67 / 0.59
		Kubo	Boiyemety (6)	[4]	0.67 / 0.71
Southern Highlands Province Subclan (Tribe) Now within Hela Province					
6 groups	8 [1.33 each]	Various[c]	10 groups (89)	[2]	10 [1.00 each]
		Various[c]	9 groups (65)	[3]	24.43 [2.71 each]

Sources: Moro agreement (Juha Petroleum Retention Licence 2 Licence Benefits Sharing Agreement) 2009 and DPE *Recognised Major Clan and Sub-clan List* 2014.

Notes: Proposed distribution to named Febi, Kubo and other clans according to the Moro agreement and the DPE *Recognised Major Clan and Sub-clan List*. The penultimate column of the table records regions as diagnosed by DPE: '1 = Fembi Western Province' (50% of benefits), '2 = Fembi Hela Province' (10%), '3 = Tuguba and others' (38%) and '4 = Other Western Sub-clan (Tribe)' (2%). With the exception of 'Southern Highland Province Sub-clan (Tribe)' the final column shows per cent allocation of benefits based on, respectively, a distribution to major clans and a distribution to subclans. All spelling of clan names is as in original documents. We have provided language attributions.

a. The Moro agreement specified benefits distribution only with respect to Coordinated Development and Operating Agreement (CDOA) equity. The Licence-Based Benefits Sharing Agreement produced for Hides 4 APDL 7 was more thorough and informative than the equivalent Moro document. The former states that the distribution of benefits to recognised landowners is the same for different categories of benefits: royalties and two forms of equity. We have assumed that this was intended to be the case in the Moro document (viewed 11 August 2014 at: actnowpng. org/sites/default/files/PRL%2012%20Hides%204.pdf).

b. The people classed as Mora appear to be the descendants of Etoro-like people who, in the early 1970s, lived on the slopes of Mount Sisa towards the headwaters of Baiya River and spoke a language that is sometimes named as Tsiani (see Chapter 3, Note 6; Denham et al. 2009: 4–23). In Ernst's (2008: 61) list of Febi clans, Amado is listed as a subclan of Mora; in the DPE list of recognised major clans and subclans Amodo is listed as a major clan. We have assumed they are the same; they may be more Huli-like than Febi-like. In the context of the PNG LNG Project, Mora and Amodo people emphasise their connections with Febi. If the latter are regarded as Huli, the benefits to Kubo and Febi would be somewhat less than 65.5 per cent.

c. The language affiliations of these named groups are mixed. The six groups named in the Moro agreement include at least one that is Duna-speaking (Yandika; Goldman and Ernst 2008: 97). Pari is considered to be a Huli-speaking group by Goldman and Ernst (2008: 97). The list of recognised major clans and subclans names ten groups from Region 2 (Fembi Hela Province) and, excluding Febi and Kubo clans, nine from Region 3 (Tuguba and others). It is possible that some of the latter represent upgrades of former Koli subclans to 'major clan' status. At the clan vetting meeting held in November 2013 the Region 2 groups asserted residential affiliation with Siabi, the village nearest to the Juha well heads. Most of these appear to be Huli-speaking groups based in Hela Province, some of whom (for example, Bebe) assert that their origins lie in the Juha region (Denham et al. 2009). Two of these ten groups–their binomial names are prefixed with Pogaiye–may be Bogaia-speaking (Goldman and Ernst 1980: 182). The groups from Region 3 asserted residential affiliation with villages named Gesesu, Sebitu, Timalia or Tokaju. Some of these are associated with land that spans Western and Hela Provinces in and near the upper reaches of the Baiya River watershed (for example, those affiliated with Sebitu) and, though primarily Huli- speaking, at least a few are strongly connected by both marriage and residence with the clans Koli and Mora. Of the 19 named groups discussed here, only one (Togumu) appears in both the Moro agreement and the list of recognised major clans and subclans. This is because previous claimants refined their own understanding of the classification of groups and because other groups were successful in claiming rights to benefits. We ourselves have no direct knowledge of any of these groups.

A copy of the DPE list arrived at Suabi on 17 April 2014, two weeks before we departed. Those who saw it, or were asked by others what was in it, had one immediate interest. They needed to know whether they, their subclan or their clan was named. Most who expected to be named were happy. When we asked about groups that were missing the usual response was 'That's their problem'. A few noticed that some Hela Province clans that they thought had been excluded at the clan vetting meeting had reappeared on the list. They didn't know why that had happened: 'perhaps bribery' was the suggestion. But in those two weeks nobody had engaged with the fact that the stated distribution of financial benefits from the Juha gas wells was totally at odds with the distribution of benefits that had been set out in the Moro agreement. Their concern was with their own eligibility as future recipients of benefits. At this time they were not alert to the 'losses' they would incur because the share to Febi and Kubo clans had been greatly decreased. Nor, yet, were they aware that the distribution among Febi and Kubo clans—among the people they lived with on a daily basis—would differ substantially if that distribution was made on the basis of equal shares to subclans within a designated region rather than equal shares to major clans within that region. In the former case, more benefits would flow to those major clans that had subdivided to the greatest extent. For example, under a distribution to major clans, Wuo Fofasoso and Wuo Tigidin would each receive 2 per cent of benefits; under a distribution to subclans the benefit to Wuo Fofasoso would increase to 2.25 per cent and that to Wuo Tigidin would decrease to 1.13 per cent (Table 5.1). Or, again, where members of the greater Headubi assemblage would share 8 per cent of benefits under a distribution to major clans, this would drop to 4.96 per cent under a distribution to subclans. At the present time these disparities exist only on paper. They will be made real, and their consequences will be felt, when, or if, money is actually forthcoming.

The environment of bureaucracy is never stable. In PNG the process of registering ILGs had become messy (Koyama 2004; Lea 2013; Weiner 2013). The Land Group Incorporation (Amendment) Act implemented in 2012 was designed to address issues of economic inefficiency and corruption that were associated with that process. To that end, the requirements for registration were made more rigorous than had been the case before. People intending to register an ILG would be now obliged to provide, among other things, a constitution, a minuted record of meetings, a formal birth certificate for all listed members and a sketch map of the land under consideration. The birth certificates were to be obtained from the Office of Register General (Civil Registry), Department of Community Development, at a cost of PGK15 per adult. The sketch map was to 'consist of a boundary walk by clan leaders, Provincial and District Lands Officers, a land Mediator and the Provincial Surveyor', clearly mark 'any disputed boundaries', be 'signed by the disputing clan elders to acknowledge the existence of the dispute' and be 'drawn on a A3 paper by the Registered Surveyor using topographical map of the area or GPS coordinates to indicate the size, location and the exact boundary' (GoPNG 2012a: 3). Across PNG, pre-existing ILGs have been terminated, and new applications that meet these criteria will have to be submitted.

People at Suabi were unaware of the revised requirements. Nor, in early 2014, did we know of them. And people would have been unable to comply with them in the form they had been enacted. No one at Suabi has had their birth formally registered and few adults know their age, let alone their date of birth. Procuring birth certificates necessitates filling in an intricate form for each person, and for an ILG comprising 150 members could cost as much as PGK2,250. An individual who appeared with different names on different lists would, presumably, need more than one birth certificate. Further, the practicalities and costs entailed in preparing a sketch map in accordance with the strictures summarised above would place the exercise beyond possibility for either local people or nominated officials. At Suabi some people have prepared written accounts of waterways, ridges and gullies that mark the border of what they consider to be the land of their clan. But these accounts have not been prepared in consultation with others who might dispute that border. Indeed, in the recent past there were no such borders. In the 1990s, Desa, who was then a young teenager, explained the geographical relationship between her natal land and that of Martin, who was of a different clan. She held her hands in front of her, placing the fingers of one hand between those of the other. 'Our lands are like this,' she said, flowing one into another. Identification with land gradually decreased as a person moved away from key sites of activity. In those years 'foci, not boundaries, inform[ed] Kubo discourse about land and, for each individual, these [were] established as salient through personal and shared experience' (Minnegal and Dwyer 1999: 66–7; see also Minnegal and Dwyer 2011a; Dwyer and Minnegal 2014). Lines drawn on maps—with their connotations of 'ownership' and 'property rights'—intrude on the lives of other people. They may be acts of appropriation, of taking to oneself (Busse and Strang 2011: 4). They can fuel disputes. The revised ILG system is likely to reverberate through the lives of Kubo and Febi people for a long time to come.

Accessing Money

At Suabi, people understood that the Juha wells were not scheduled for production until 2020. Only then would they begin to receive massive royalties for the gas from those wells. In the interim, however, it was also their understanding that some royalties would be paid to all landowners involved with the PNG LNG Project from June 2014, when the Hides field was to be brought into production and the first gas shipped overseas. Later, when Juha came on line, those advance payments would have to be reciprocated, but for the present that pending constraint could be ignored. The first shipment of gas took place, ahead of schedule, in May 2014. But the Bank of PNG Governor, Loi Bakani, cautioned stakeholders, including the government, not to 'expect a windfall of revenue in the first year of production', warning that 'Export receipts and revenue for the government may be minimal and therefore expectations of windfall revenue and any associated appreciation of the kina might not materialise in the near term' (Sharma 2014). Prime Minister Peter O'Neill said that it would be 2015 before money from the 'multi-billion Liquefied Natural Gas' project was available (PNG Today 2014).[8]

Though royalty payments would come in the future, there was other money set aside by government that was, officially, already available—PGK1.2 billion as Infrastructure Development Grants and PGK120 million as Business Development Grants, with PGK11 million of the latter earmarked to assist in the development of local business ventures within the Juha licence area (see Chapter 3). At Suabi many people, or groups of people, sought funding through both categories of grant. Often they brought their applications to us to be typed or discussed. Infrastructure grants were available to communities for projects such as school buildings and health centres, airstrip construction or maintenance

---

8   Through the early months of 2014, people at Suabi were under the impression that government officers would either come to the village or call them to Kiunga and assist with the registration of Incorporated Land Groups. Such assistance had been guaranteed under the Moro agreement (see Chapter 3) but had not happened by December 2016. Nor, to that time, had people received any royalty payments; they had been told at different times that payments were on hold because global oil and gas prices had collapsed or because clan vetting had not been finalised.

and upgrading of walking tracks that linked well separated villages. Business Development Grants were available to assist locally registered companies start up small businesses.

Between May 2005 and May 2014, some Febi and Kubo people were named in association with at least 49 Business Names, Business Groups, Associations and Companies registered with the PNG Investment Promotion Authority. Registrations peaked in 2010 (n = 17) and 2014 (n = 9 before May) following, respectively, the Licence-Based Benefits Sharing Agreement held at Moro and the clan vetting meeting held at Siabi.[9] Febi people were more likely to be named as company directors than Kubo people (45 of 54 unambiguous listings) and most registrations included the names of people other than Kubo or Febi. These names were often those of lawyers, accountants or others (including Huli men) who had been contracted to assist with the registration process. In some cases the person who had provided assistance was the only named shareholder. With the exception of Bob Resa's companies we know of only one other case where a local landowner company was awarded 'seed funds' and, here, as in Bob's case, the key recipient used the grant to relocate to Port Moresby. During the years 2011 to 2014, some people who lived at Suabi felt that they had been cheated by those who received money and departed. Most, however, felt that they themselves must have erred, that they had not gone about the process of accessing funds in the right way. They wanted to start again and, where possible, do so with our guidance. In what follows, we describe people's actions through the period we visited Suabi.

Infrastructure Development Grants

By December 2013, at Suabi, one infrastructure grant of PGK81,500 had been received to support maintenance work on the airstrip and fund the building of a new community health centre. Several thousand kina of this money 'disappeared' and was rumoured to have been misappropriated. For several months, at the market and at a concluding court case convened for the purpose, there was a succession of accusations and counter accusations and, on one occasion, men came to blows. The court case—held at an open-sided shelter beside the Council's house over two days, overseen by

---

9   We searched the PNG Investment Promotion Authority website (www.ipa.gov.pg) for registrations that included likely language, clan, village and other place names allowing for multiple variations in spelling.

the Law and Order Committee (the Council was one of those suspected of misusing the funds), and attended by a floating audience of men, women and children from all Corners—eventually provided a public resolution to the matter, though it left some people unsatisfied. Many people spoke. Details of expenses were written on a large board for everyone to see, receipts were presented and the money was 'traced' from its origin point—its arrival at Suabi—to its current endpoint. Once the 'path' of the money had been established, the matter was judged to have been resolved. A small payment was made as compensation to a man who had been falsely accused of misappropriating money. There was no discussion of the wisdom of how the money had been used. Some had been spent on personal needs by men who were tasked with buying building materials in Kiunga. Those men regarded the money as appropriate exchange for the services they were rendering. To bring materials from Kiunga a plane had been chartered at a cost of PGK5,000. Three drums of fuel, at PGK1,013 each, were included in the cargo, intended to run a walk-about sawmill that had been purchased more than six years earlier and was currently stored in a privately owned container. It was only after the fuel arrived that the container was opened and condition of the sawmill checked. There was no battery. New spark plugs, blade, tyres and filters were needed and all parts were thick with rust. A mechanic from Juha inspected it and judged it to be inoperable. An earlier inspection, before the fuel was bought and a plane chartered, might have encouraged different decisions about how the money might be best spent. But earlier inspection would have been counter to Kubo practices; these people are very disinclined to maintain equipment in advance of a need to use it. Nor had an additional complication been resolved. The man who had been nominated as the sawmill caretaker—the owner now lived in Port Moresby—had set rent at PGK600 per day.

Three other applications for infrastructure grants were prepared in the next few months. All sought money to fund new community school buildings. All stressed that Suabi was a remote community receiving little government support, and that current school buildings had been made of bush materials and were now in very poor condition. All three applications sought funds for a single classroom, a double classroom with central office and two teachers' houses. They were independent submissions. One sought PGK150,000, one sought PGK450,000 and the third sought PGK2,000,000. Neither the first nor the second provided any rationale for what was requested or provided justification for the amount of money requested. The third, however, was exceptionally detailed. It originated from the combined efforts of one Suabi resident and friends of his who were the directors of a Goroka-based company. There was little doubt that the Goroka friends were more interested in feathering their own nest than in providing school buildings for a remote village they had never visited.

Esau had not yet drawn up applications for infrastructure grants. But that is what he intended and his preliminary efforts entailed a grand vision. He had recently been appointed Council to an area that included most of Febi territory. On a visit to Goroka, in the Highlands, he had paid someone to draw up a five-year plan for the Ward that he oversaw. The document stressed, accurately, that there had been no development in the area since Independence; there were no roads, no schools, no health centres, no government-based law and order system and there was no cash cropping. Esau's five-year plan was to change all this. It envisaged a vehicular road from Nomad to the Juha area and three towns, each with airstrip, schools, community health centre, streets and shops, emerging where currently there were none. Cash cropping—in particular, vanilla— would be developed.[10] The document named the places where these towns would be developed, and assigned to each starting populations of between 300 and 800 people. Esau agreed with us that these numbers were very high—in fact no one lived at one of the three places—but said that he had included all people who had expressed interest in those places. To emphasise the remoteness of these locations their distances from Nomad and Kiunga were recorded, and exaggerated by a factor of 100. And, for each of the three selected town sites, in tabular form, the document provided breakdowns by sex and age and summaries of major causes of death and hospitalisation. (All numbers were fabrications and the percentages recorded for deaths and hospitalisation tallied to more than 100.) Another table listed law and order problems, with violence directed against women ranking highest and drug use considered significant.

---

10   The desirability of establishing new villages or towns, similar to Suabi or larger, was mentioned at different times by a number of people. People were concerned that the land associated with many clans was seldom visited and understood that, in both their own customary practice and government expectations, ownership rights over land that was not used could disappear. They argued that as local population size increased people would be able to return to their own clan lands, establish large communities, and continue to enjoy the advantages that currently existed at Suabi.

Neither applies to the Febi and Kubo populations.[11] Esau's document was strongly influenced by the understandings of the Highlanders who helped him prepare it.

Business Development Grants

Access to Business Development Grants was more difficult than access to infrastructure grants, because only those with a registered company were eligible to apply. In earlier years some people had submitted applications through Bob Resa's company; none had received money, though some suspected funds had been forthcoming and not passed on. Few were now willing to submit applications through that company, despite considerable pressure from Bob to do so. Instead, people now expressed a need to establish and register their own companies. But they were unsure how to go about this, and the majority did not appreciate that registration of a company and application for a Business Development Grant were distinct procedures handled by different arms of government.[12] Thus, for example, three women's groups drafted letters in which they asked, in the one document, that their proposed company be registered and that they be provided with sewing machines, rolls of cloth, needles and cotton to make clothes to sell at local markets.[13] At least these women knew what kind of business they wished to initiate and, wisely, were modest with

---

11   There is very little violence against women at Suabi. In seven and a half months we knew of one fight between husband and wife and two occasions when a man hit his wife when he learned that she had had an adulterous affair. Concerns about young people—'school boys'—drinking and taking drugs (marijuana) was a common theme in speeches at the market though, in fact, there was a low incidence of both. Some alcohol was smuggled into the village on Mission Aviation flights and some home brew was made. On one occasion several youths behaved raucously after consuming home brew, and one, who was very drunk, shot an arrow at his father after being ordered to stop. On another occasion, at Kiunga, a Suabi resident was jailed for fighting with people, including police, while drunk. This was a source of much gossip and, for the man concerned, deep shame when he eventually returned to his home village.

12   The PNG *Companies Act 1997*, Section 13(1)(a), states that persons seeking to register a company must provide an address that includes the suburb, street name and number of the registered office and, where the address is an office, 'particulars of the location within the building'. Neither a post office box number nor the name of a village is acceptable as a legitimate address. People living in remote areas of PNG can comply with this requirement only by establishing a relationship of indebtedness to a lawyer or other person who acts on their behalf with respect to providing that address.

13   In these and some other cases, the proposed corporation was termed a 'Business Group' and not a 'Company'. Under PNG legislation, however, 'Business Groups' and 'Companies' differ with respect to rights, obligations and registration procedures, with the latter subject to more constraints than the former (GoPNG 2014a, 2014b; Kalinoe 2003; Jackson 2015). At Suabi, people did not understand these differences. They were under the misapprehension that a legally registered 'Business Group' would be eligible to access Business Development funds.

respect to its size. Others were quite unsure what kind of business they might establish. One man had obtained a government document that outlined the sorts of businesses that could be eligible to receive funds. His subsequent application to register a company included rice cultivation, livestock husbandry, inland fisheries, production of minor crops (vanilla, chilli, cardamom), market gardening, bakeries, artefact sales, trade stores and logging as the kinds of businesses that the company would undertake. He simply listed all those offered as examples on the guide sheet.

The eight applications that we saw (though none had been submitted by mid-2014) varied in the detail provided with respect to the planned business, but were consistent in failing to provide justification for the funds requested. One well constructed application, from a Kubo community to the south of Suabi, sought PGK250,000 to establish farming activities (crops, livestock and aquaculture), sawmilling and a tourist lodge that would take advantage of horticultural and forest-growing orchids. The application outlined preliminary efforts and local competencies in some of these areas and asserted that their viability would be enhanced by the Trans Papuan Highway that the applicants expected to reach Nomad. The application itemised material goods that would be needed to fulfil some of the stated objectives but provided no costings. Some people asked for 'seed money'—this approach had scored a satisfactory outcome when used a few years earlier—but with no clear indication of what that money would 'seed'. When we asked one comparatively well educated and literate man what he would do with the money if his application were successful, he responded simply that he would put it in his bank account.

Money that had been allocated by government to fund Business Development Grants was understood by local people as a potential resource that was desirable but difficult to access. There was little prior experience, and few guidelines, with respect to access mechanisms. At face value our examples above are discouraging, implying incompetence on the part of applicants. But they may be also understood in a different light. Garnering resources from the environment always entails use of appropriate techniques. If the category of resource is new then new techniques are called for. At Suabi, people were experimenting with possible techniques. Some acknowledged that they were 'just trying', that if what they were currently doing failed then they would try something else. That is what they had always done. That is why, in the 2000s, they switched from hunting and trapping wild pigs to snaring them; the latter technique proved more efficient in the new context of the large number of people, and their impact on forest land, at Suabi. Once resources have been appropriated, however, they belong to the person who appropriated them or to the people to whom he or she is in relationships of debt, trust and responsibility. It is this collective of people who decide how the resource will be distributed and used. Those judgements are not made prior to the act of appropriation. People at Suabi were, therefore, confronted by a dilemma. It was necessary to employ complex techniques that they did not understand to garner funds under the Business Development Grant program. But only if, and when, they were successful would it be time to decide how the allocated funds should be distributed and used. From their perspective, the procedures they were obliged to follow placed the cart before the horse; they were required to state, on paper at least, what they would do with the money before they had received it. For this reason, at least in part, there was more secrecy with respect to these intended transactions than in other financial domains. As people moved towards finalising applications they became anxious that others might see their documentation and, as they expressed it, 'steal points'. For a few people that anxiety intensified and spilled over as depression.

Navigating the Future

When we arrived at Suabi in December 2013, one of the men we expected to see and wanted to talk with was absent. Urie was in the Highlands. He had gone to Goroka, we were told, to 'buy a wife'. The news puzzled us. Urie was a Febi man but not from the group of clans that form Wuo. He had worked at Juha for several months before the November clan vetting meeting. Through that time he would have earned, perhaps, PGK5,000; enough for travel and support in the Highlands but by no means enough to pay bridewealth for a Highlands woman. Moreover, he had seemed content in his marriage and his oldest child was in his twenties.

It was early February before Urie returned to Suabi. He came to see us three days later, and again on several subsequent occasions. We talked for hours. He was depressed, overwhelmed by feelings of ignorance, not knowing where to turn for help. In November, at the Siabi clan vetting meeting, he realised that men he had regarded as close kin were sidelining him. His mother's first husband was a Wuo man, and he regarded her son by that man to be his brother. But his 'brother' now excluded him from ILG lists and from proposals to register companies and seek grants for developing businesses. Senior Wuo men were treating Urie's clan as having few, if any, rights to benefits that would flow from the Juha wells. Urie decided, therefore, that he would have to find his own 'road' to secure the future of his clan. But he was unsure how to go about this, felt he needed the guidance of a '*savi man*'—a man of knowledge. He would like this to be a white man or white woman who understood the intricacies entailed, knew how to negotiate them, and would help without charge. But the only white men or women he knew—people like us—came and went, and always had their own agendas. We asked about his son Madison who, it was said, had completed Grade 12 and been accepted to study at university in Port Moresby. 'He is a student,' Urie told us. 'He is not a *savi man*'. Young men of Madison's generation were inexperienced, lacked the maturity that comes with marriage, and were too often fickle or self-interested in contributing to such serious endeavours. And, though Urie did not say it, he did not want to be beholden to those who were a generation younger than he was.

So Urie travelled to the Highlands where, he knew, there were many Papuan New Guineans well versed and experienced in the bureaucratic procedures he needed to navigate. He bypassed Tari, where Huli-speakers angered by the outcome of the Siabi clan vetting meeting had threatened harm to any Febi or Kubo people who passed through their land. Indeed, for the first few months of 2014, people at Suabi considered that Tari was too dangerous to visit and a few school students made alternative arrangements for that year. Urie travelled further east, to Goroka, where he would be distant from Huli influence. He bought a return air ticket to ensure he could get back to Suabi when his money ran out. And he found 'knowledgeable' people who he thought might be able to help. He presented himself as a Juha landowner, as a person who, very soon, would receive massive royalties from the PNG LNG Project on 'his land'. He was introduced to a young woman as a prospective future wife. If he married her and established affinal connections with her kin then they would be bound together in relationships of mutual trust and responsibility. It would potentially be a 'win-win' outcome. But nothing was finalised. His prospective affines wanted to see and receive the money before they would do what Urie wanted. And, as yet, there was no money. As yet, Urie could not even be sure that his name and that of his clan would be listed by DPE as legitimate beneficiaries. Urie had spent all the money he had to put a possibility in place, a possibility that, as is always the case with the future, remained deeply uncertain. He had embodied that uncertainty and was depressed.

Kubo have always lived in a world of uncertainty. There is a strong sense that they are agents in shaping their future—that who they are, or might be, is not determined by what they, or others, have done in the past but, rather, is shaped (and revealed) by their own actions in the present. Their focus is on recognising, and acting on, the ever-changing opportunities that emerge in the world. Always, however, this is done in the knowledge that outcomes may not be as hoped; other agents, too—human and non-human—are pursuing their own agendas, and this may alter the consequences of action.

Through the past decade, Kubo have glimpsed new possibilities, become tantalised by hopes and desires for a very different future. As before, they are acting on the opportunities they see in the present to bring that desired future into being. But the mechanisms entailed have not yet been mastered. There are agents at play now who they do not know, yet whose decisions have profound effects on both the opportunities available and what must be done to seize those. Possible paths to the future are ever-changing, as legislation alters, bureaucracy brings in new rules, and others pre-empt moves. And the 'destination' has been no more than glimpsed. As Kubo strive to move into the future, the world moves around them, shifting currents and waves that converge on Suabi often from beyond the horizon. Thus, as Vigh wrote:

> Beacons translate … into hope (or fear)—not a given, durable point in a 'scape', but a somewhat frail and delicate imagined position in the yet to come, which changes in relation to both the movement of the agent and of the social environment (Vigh 2009: 433).

People at Suabi continue to seek new 'roads' towards that beacon, continue to navigate imagined future possibilities. In the process, they are transforming their own social world, bringing into being new collectivities, and new kinds of collectivities, that will dramatically alter potential actions for each of them. But that same process has, paradoxically, been grounded in a revisiting—and re-visioning—of past actions and interactions.

6. Navigating the Past

Gwia died in March 2008, of severe anaemia. He was probably in his early forties at the time. About 20 years earlier he had married Wuagodua. Their first two children died but, at the time of Gwia's death, there were four living children—two boys and two girls—and Wuagodua was four months pregnant. Gwia 'did not see the face' of Wuagodua's last-born daughter, Jorah.

The marriage between Gwia and Wuagodua had not been reciprocated. There had been no exchange marriage that balanced their union, nor a payment of money or pigs that would serve in its place. Gwia's clan affiliations had been with Demeti, Wuagodua's with Yawuasoso. In lieu of an exchange marriage, Yawuasoso men had initially been given Jano, the second-youngest daughter of Wuagodua and Gwia. She was to become a 'sister' to those men and, later, be the exchange bride for one them when he married. Now, however, Yawuasoso laid claim to all the living children who had been born to Wuagodua before Gwia died. (They asserted no claim to Jorah because, having not been seen by her biological father, she was not considered to be properly of his clan; there was no need to claim her.) But were their claim to those children successful, Demeti would be left with no male representatives. Demeti was one of seven subclans within a greater assemblage—a major clan—known as Dobiti. As the oldest female representative of that larger assemblage, and the only living representative of one of the subgroups, Bosofi was concerned that Demeti might cease to exist. She lobbied her kin. They should give money and a pig to Yawuasoso to legitimise Gwia's marriage, and thus retain their rights over his children (Fig. 6.1).

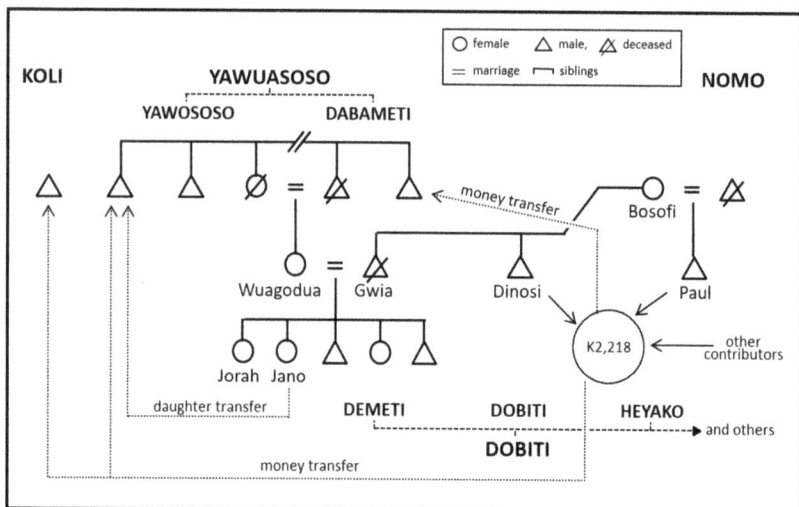

Figure 6.1: Primary participants in a payment of compensation for a failed marriage exchange.

Source: Authors.

Note: Major and minor 'clan' (or 'subclan') names are in bold capitals.

In February 2014, people assembled for the exchange. A large pig, provided by Bosofi, was tethered to a post and, after much manoeuvring back and forth, seeking position for a fatal shot, it was dispatched by one of Wuagodua's clan brothers. It took some time to die. Dogs gathered as it thrashed on the ground. Martin blocked his ears; he did not want to hear the pig 'scream'. Tinus nursed a small boy—Javan, the son of Dinosi—who had not before seen a pig being killed, and who watched in wide-eyed horror, begging his carer not to leave him.

The carcass of the pig was displayed in front of four bamboo poles, three of which were festooned with money that totalled PGK2,218 (Fig. 6.2).[1] The pig was valued at PGK1,500.[2] There were many speeches, by both men and women. The money and the pig were now being given to

---

1   There was no occasion, in the years 1986 to 1999, when we observed money to be displayed in the course of an exchange (Minnegal and Dwyer 2007: 8, n. 7). However, Ernst (2008: 63–4) illustrates the public display of money as bride wealth among Febi people in 2008.

2   The monetary value placed on pigs depends on several different criteria. Domestic pigs are valued more highly than wild pigs of the same size, reflecting the work entailed in rearing the former. Butchered portions of wild pig—legs or shoulder—that reach Suabi from Strickland River communities sell at higher prices than equivalent portions hunted locally, reflecting the effort entailed in carrying the meat. Further, a domestic pig that is sold live is valued more highly than one of the same size that has been killed and butchered before sale and, though no money actually changes hands, a domestic pig that is included in a compensation payment is valued even more highly as an explicit expression of its 'symbolic capital'.

Yawuasoso in compensation for the failed exchange of 20 years earlier. Gwia's children would retain their Demeti heritage. But there was one exception. One Yawuasoso man, it was said, had misbehaved. His antisocial acts were not enunciated but everyone knew that he had raped a local woman and shot a white man. The community was shamed by the trouble he had caused. Gwia's brothers insisted that they would give this man no money. That is why one bamboo pole was bare. In lieu of money, it was declared, Yawuasoso could keep the girl Jano, the second youngest of Wuagodua's children.

It was time now to claim the money. A senior Yawuasoso man—Wuagodua's classificatory father—took two poles that together displayed PGK1,218. A senior Koli man—his mother's sister was mother to one of the Yawuasoso men—took the third of the poles that displayed money. He received PGK1,000. Then there was a pause. People waited. Would the empty pole be accepted? Would it be acknowledged that the man for whom it was marked—he was absent from the village—had been publicly declared *persona non grata*. Eventually, the man's brother stepped forward and took the pole. The exchange was finalised. The future of Demeti had been secured.

The exchange of money and children described above brought closure to a long-standing dispute that had been the subject of negotiation at several recent locally convened court cases. A long-standing failure with respect to balancing a marriage had now been set to rights. The pig reinforced the sense of closure. It was cooked, butchered and widely distributed through the late afternoon on the day of the exchange. The sharing of its meat served to generate goodwill within the greater Suabi community, a community where a sense of camaraderie had been disrupted by the fact that past failures by some people had resurfaced in the present.

Why, though, had people waited 20 years to initiate and finalise this payment? In large part, it was money that was at stake. Or, rather, it was the perception of inequality generated by the fact that access to money was unequally distributed. Some people were employed, most were not. Two men—Bosofi's son Paul, and the senior representative of the nominal clan Dobiti, Dinosi—had, through the past two months, worked for petroleum companies: the first as a trainee helicopter loadmaster, the second as a security officer for the Talisman campsite. Both were known to have been paid more than PGK1,500. Both were known to have obligations that they could not avoid without loss of face; Paul to his mother Bosofi, and Dinosi to all the members of the greater Dobiti assemblage. These two men contributed most of the money that changed hands that day.

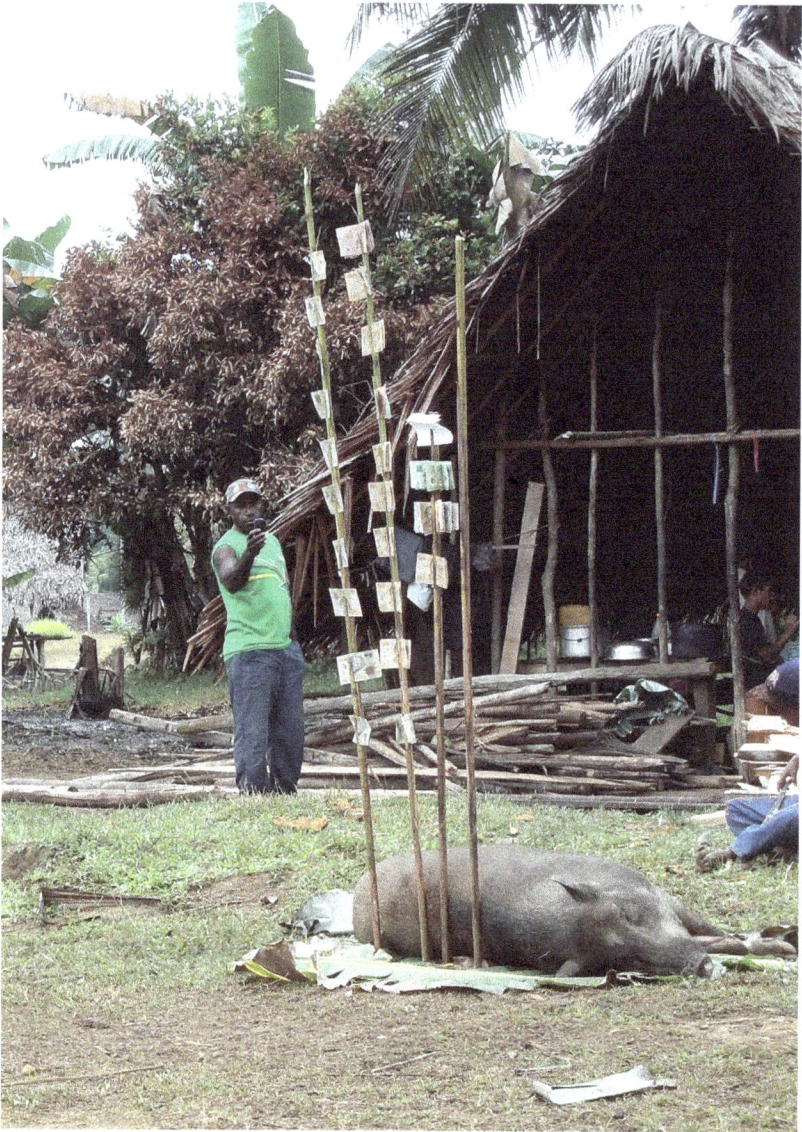

Figure 6.2: Exchange of money and a pig in compensation for failure to reciprocate a marriage.

Source: Photograph by Peter D. Dwyer, 2014.

The revisiting of that long-past failure was a response, then, to shifting realities in the present—shifts that had little to do with the actions of Gwia and his wife 20 years before. As people at Suabi seek to position themselves in relation to the opportunities offered by a world increasingly shaped by global capital flows, they are selectively mobilising connections established long before. They will have always done so. But, in drawing on a past that played out in a very different world, they are rereading the events and understandings that shaped that past. At Suabi in 2014, this played out not only in the resolution of disputes through the new mechanism of village courts, but also in the reshaping of mythological pasts that accompanied the differentiations and amalgamations discussed in Chapter 5. This chapter focuses on these new ways of validating action in the present by selectively drawing on the past—by navigating a past that could always be read otherwise. The readings now being foregrounded, however, project into that past what are new interests and imperatives.

Courting Impropriety

Court cases were a regular feature of life at Suabi. The initiating causes were diverse: premarital sex, adultery, failed or unsatisfactory marriage exchanges, unresolved separations of husband and wife, repeated public fighting by a husband and wife, ongoing disputes between co-wives, fights between men, disputes over land, sorcery accusations, damage to gardens caused by a domestic pig, or concerns about the way in which money granted to the community had been spent, all featured as the basis for a court case. They could be initiated by a single offended party, by a group of people or by the community as a whole. They were adjudicated by the Suabi Councillor if he was in residence. In his absence, or when he himself was a protagonist, one of the men appointed as either 'Law and Order Committee' or 'Corner Committee' took the leading role. And, in their absence, other senior men chaired the court case. The cases might be resolved quickly, but more often, took many hours in a day and sometimes needed to be reconvened on a later occasion. They were public events. Everyone could come—men, women and children—and many did.

At most court cases, two styles of evidence were foregrounded. The first concerned the present—what had been seen and heard—and the second concerned the past—what was appropriate or inappropriate relative to past practices and present expectations. The following brief examples provide illustrations. We include cases that we witnessed, others that we were told about and some that are pending.

The Gwia court case described above foregrounded relationships that, in the past, had not been appropriately balanced. But, in the way it played out on a public stage, those who asserted offence displayed more concern for compensation than for restoration. Justice was to be measured in terms of money. Indeed, this was accepted by those who paid, for in refusing one claimant they effectively placed greater symbolic value on money than on the girl they handed over in lieu of money. Here, then, a resurrected past failure in relationships was resolved by means of the anonymising properties of money. Other cases reveal a similar approach. The first of the following three examples concerns inappropriate disposal of the dead, the second a failure to consult as expected, and the third an accusation of sorcery.

> Kokobe's husband Imoi was in the forest, well away from Suabi, when he died. People who were with him at the time carried the body to Suabi and placed it in a house. Imoi had died on the land of Dobiti people and it was men of that clan who, under customary practice, should have carried the body. But there were no Dobiti men present. The men who carried the body were of the wrong clans and, further, offended against convention by placing the body in the wrong house. A court case was convened and, for their failures with respect to custom, three senior Dobiti men incurred a joint fine of PGK800. The fact that present-day living arrangements, with few people resident in the forest and with travel to more distant locations common, made it increasingly improbable that customary practice could be followed had no bearing on the judgement. The men themselves accepted the finding of the court and, as a gesture of goodwill and enhanced community relations, increased their payment by PGK300.

> In 2012, Noima took a second wife. He was an Awasoso man and his new wife—a Samo woman—was a widow. Her Yawuasoso husband had died about ten years earlier and, at his death, Elei had returned to her home village. Noima negotiated his marriage with her kin and not with her first husband's kin. Biko was enraged. He was the most senior living Yawuasoso man and, though he was not from the same branch of that clan as the deceased man, considered that he should have been consulted. He collected his bow and arrows, ran to Owabi and fired at Noima. The arrow missed—as Biko probably intended that it should. He was publicly declaring that Noima had given serious offence and laying the groundwork for future compensation.

> Uwago was very ill. A few years earlier, while gardening, she slipped and was seriously spiked in the groin by a length of wood. For a year or more she was house bound. When she needed the toilet she had to be carried

there on a stretcher. Now she had tuberculosis and was often a patient at the community health centre. On one such occasion, she woke in the morning to discover that she had been 'turned around' during the night; her feet were where her head had been. And she had dreamed of being visited by a man. It was this man who had turned her and, she asserted, was responsible for her illness. He had ensorcelled her. On that night, the elderly woman Ogu had also slept at the health centre. Thirty years earlier, Ogu had been married to Okaibo but the marriage was short-lived. But Ogu's presence that night provided the link Uwago needed. She declared that it was Okaibo, acting in relation to his failed marriage, who had come to the health centre and turned her in her sleep as part of his ensorcelling technique. With her son she called for a court case. She presented her argument and sought monetary compensation. Others reported that doctors in Kiunga had declared that Uwago's illness could not be cured, and that a Nomad-based spirit medium had failed to implicate Okaibo. The court concluded that Uwago's case was based on mere suspicion; she had not actually seen Okaibo do anything. She had accused him unjustly. It was Uwago, and not Okaibo, who was required to make a small compensatory payment.

Sexual Transgressions

In various ways the cases summarised above either drew on the past for models that revealed 'proper' practice, or allocated responsibility for inappropriate behaviour on the basis of relationships—*oobi* identity, siblingship, marriage failure—that could be traced back into the past. The next three cases concern current relationships and seek resolution through an appeal to a more general sense of what constituted appropriate behaviour in the present. These cases were convened to investigate charges related to premarital sex or adultery. While such behaviours disrupted particular relationships, they had more far-reaching implications for the greater community. Here, then, though practices drawn from the distant past did not feature in the evidence and the arguments brought forward, the sense that community cohesion was central to resolution underlay judgements reached and penalties imposed. And irrespective of what was said at the cases—indeed, also of what was not said—expressions of community cohesion were irrevocably linked to the ethos of earlier times.

Here then, we offer one example of a dispute about premarital sex and two of disputes that asserted adultery. If premarital sexual relations led to marriage—an offer that was always put to the accused couple—discussion turned to the requirement that the marriage should be either balanced by exchange or entail bridewealth. Where adultery was involved, however, negotiations to seek a resolution were again explicitly framed in monetary terms.

Bimadua arrived at her parents' house at dawn. She had been *in loco parentis* to Awai and, the previous night, Awai and Bimadua's brother Jason had slept together. People gathered as Bimadua, standing outside the house, told what had happened. Jason stood to one side, head bowed and silent. A court case was convened at once. Jason was 'afraid to go to court'. Some older youths and young men went to speak for him. Awai's parents demanded that the pair marry, but neither Awai nor Jason wanted this. Jason wanted to return to the Highlands and complete his secondary schooling and, anyway, his parents refused to help with bridewealth. Their son was being troublesome, they said; he was not contributing to the needs of his family, was not behaving as a son should, and thus did not deserve their assistance. His sister Bimadua, too, refused to help. By having sex with the girl while she was staying in his sister's house, Jason had transgressed the implied 'sibling' relationship between them. Jason's supporters argued against marriage. They revealed that Awai had 'many boyfriends', and proposed that in lieu of marriage both Jason and Awai should be fined PGK250. Both, they asserted, were to blame for what had happened. The court agreed to this proposal but Awai's parents said that they did not have this much money. Instead, they offered a pig that was judged to be of the same value. Jason's parents were obliged to follow suit.

Nick was an absent husband, more often at Kiunga or Port Moresby than at Suabi. He had left his wife and children behind. Sheri was working as 'laundry girl' at the Company camp but known to be having an affair with Kiovi. Nick learned of the affair, arrived by aeroplane from Kiunga, obtained a bushknife, located Kiovi and gave chase. He was intent on causing injury. Kiovi panicked and ran. People followed, watching and anxious. Nick's classificatory brother Zavia approached from behind, put his hand on Nick's shoulder, avoided a swing of the machete and, briefly, calmed him down. Kiovi disappeared from view. Nick grabbed a burning stick and set fire to the sago-thatch roof of Kiovi's mother's kitchen house. People tore the burning sections down and saved the building. It was several days before a court case could be convened, as people waited for Kiovi's father to return from Kiunga. The presentation of evidence revealed that another young man, Salia, had been also having an affair with Sheri. Both men were required to pay Nick PGK1,000. With help from kin, and by borrowing from others, Kiovi produced the money. Salia handed over PGK400 and acknowledged his debt for the remainder. Sheri was required to pay for Nick's return flight to Kiunga. But Nick also demanded a pig as part of his compensation payment. Kiovi's father resisted this demand on the grounds that Nick had attempted to burn his wife's kitchen house. But Nick's claim prevailed and, though thoroughly annoyed, Kiovi's mother sacrificed a large pig to finalise the arrangements. A few days later Nick departed, taking Sheri and their children with him.

Soon after Gowame left her husband and returned from Kiunga with their youngest child, it was rumoured that she had initiated an affair with Miwa, an unmarried man. The news reached her husband, Michael. He took time off from his long-term employment and came to Suabi. Court was convened. Was it true that Gowame had slept alone in her own house rather than with other women in a nearby family house? Where was Miwa late at night, when he was not sleeping in his father's house? Who was the man glimpsed by the river-bank at dawn? Was it Miwa, returning to his own home after their tryst? But nothing was established for certain. There were faults in Gowame's behaviour, all agreed; she should not have slept alone. After hours of discussion, however, the court concluded that there was no clear evidence of adultery, and that the case had been initiated on the basis of rumour alone. No action was taken against either Gowame or Miwa. The senior men who heard the case said that people in the community should cease spreading malicious gossip.

Marriage Exchanges

In the Gwia court case, and in some others summarised above, what had happened or not happened in the past served to rationalise an argument about what should happen in the present. One case, however, reversed this logic in refusing to concede responsibility for events in the past—events that in earlier days had had no significance but that now, in mimicry of other people's practices, had intruded upon local lifeways.

Martin and Tabua married in the late 1990s. At that time Martin could not provide an exchange sister to balance his marriage and, instead, paid PGK500 as bridewealth. In 2011, however, Tabua's classificatory brother asserted that Martin's payment had been merely a 'payment-in-part'. He directed attention to recent marriages where it was accepted that payments of several thousand kina were in order. Martin, he argued, had never fulfilled his obligations to Tabua's kin. His argument was supported by Tabua's stepfather, who considered that if further money was forthcoming then some of it was rightfully his. At the court case that ensued, the back-dated bridewealth was set at PGK4,000 with Martin allowed to pay in instalments. Fifteen years of inflation was ignored; present-day expectations were judged to be retrospectively valid and, hence, taken as evidence that Martin's initial payment could not have been other than a 'down payment'. Martin accepted this judgement.

But he rejected an additional claim based on a recently adopted practice among Febi, which they assert derives from the Highlands and which some Kubo people have argued they should also follow. This entails a premarital payment of some hundreds of kina to the prospective bride's mother. The payment is referred to as 'blood money', and variously interpreted as compensation for either the loss of blood experienced by the mother at the time she gave birth or the effort entailed in caring for and rearing her daughter. It was suggested that Martin should give Tabua's mother at least PGK200 as 'blood money'. He refused, partly on the basis that he himself was not the cause of Duko losing blood when Tabua was born and partly on the basis of his strong allegiance to the Evangelical Church of Papua New Guinea (ECPNG). This new practice, he told us, was not Christian. It had been taken up by those who were aligned with the Christian Brethren Church but was not supported by ECPNG.

In the case of the improper treatment of Imoi's body, people invoked customary practice in asserting their claim to monetary compensation. Invoking customary practice or, rather, an idealised view of customary practice was common in cases of adultery and those arising from disputes between husband and wife or co-wives. The antisocial behaviour was said to be at odds with the way things were in the past and, more often than not, it was a woman who was judged to have misbehaved. She had not behaved as 'a good wife' should. This remained the case even where the woman initiated an affair when her husband had been absent in Port Moresby for a year or more and was widely known to engage in sexual relations with other women. The distance entailed meant that his behaviour was irrelevant to local social relations. The behaviour of his wife in their home community was quite a different matter. It was not always the case, however, that women were the losers in these matters. Young women retained considerable autonomy with respect to accepting or rejecting suggestions that they marry particular men. Both Irene and Kalyn refused to marry men who had initiated favourable negotiations with their parents. They had no interest in assuming the role of second wife, in Kalyn's case combined with relocation to Port Moresby, and their refusals were not challenged, though, in Irene's case, a compensatory payment was made to the rejected prospective husband.

In the cases discussed to this point, information flowed freely. Many people reported details of what they personally had observed or heard, or presented interpretations of those observations. Others might question the validity of the evidence that was brought forward, and what was presented as evidence could at times be judged as no more than malicious gossip. But suppressing information was sometimes an alternative strategy for influencing judgements, though it required skill. The court process needed to be manipulated by one protagonist or another to minimise the likelihood that alternative interpretations would be forthcoming. This was particularly evident when Clem's intended marriage plans failed to come to fruition.

Clem was a widower, the father of five children who he had effectively abandoned. He was employed at Juha, earning money, and wanted to remarry. His local reputation was not good, however, so he looked further afield for a wife. He gave a young man, Leo, a few thousand kina, asking him to find a potential wife among families that Leo had come to know when he attended school in the Highlands. The money would cover travel and living expenses for Leo and allow for an initial small gift—'blood money'—to the prospective bride's mother. Leo was to report that Clem was a Juha landowner—rather stretching the definition—and, before long, would be in receipt of royalty payments. Bridewealth would then be paid. Leo was away for three weeks. He returned with Epe, a Highlands man who had supported him in past years, and Epe's fifteen-year-old sister-in-law, Tuppy, as the prospective bride.

Clem was still at Juha when Epe and Tuppy arrived. It was two weeks before he returned. Tuppy asked many questions. What was Clem like, she wanted to know. Was he a young man like Leo? He wasn't. She was not told about the five children, or the fact that Clem had sometimes struck his former wife. When Clem finally returned she was not impressed. But Clem was in a hurry; he was to return to Juha within a week. He set the day for the wedding and on that day purchased, killed and butchered a pig for a wedding feast. Everything was prepared. And then Tuppy announced that she did not want to marry Clem.

Clem was angry. A court case was convened. The adjudicators were unwilling to push Tuppy to do what she clearly did not want to do. She had the right to decide, they declared. She was given two alternatives to going ahead with the marriage: with her brother-in-law Epe she could raise the money—PGK4,652—to refund Clem's expenses, or she could marry someone else in the Suabi community. Clem refused to accept the second option. Either Tuppy marry him, he insisted, or she must repay his money and leave Suabi. If she would not marry him then he did not want to see her face.

Neither Tuppy nor Epe had the money that was demanded. Nor could they raise it without returning to the Highlands. But they weren't trusted to do the latter. They were stuck in Suabi.

Two weeks later there was a second court case, this time called by Leo's mother, Waiyo. It was necessary, she argued, to resolve the problem. If Tuppy would not marry Clem, and would not marry someone else in the local community, then for everyone's peace of mind she should leave Suabi as soon as possible. Waiyo was very forceful. Other women, too, were very much to the fore in presenting arguments. They implied that the problem had been caused by Clem's desires and haste. He had not gone about seeking a wife in an appropriate manner. He himself should have gone to the Highlands, rather than delegating Leo. Clem, once more, stated his case: Tuppy should marry him or refund his expenses and go. Tuppy was asked, yet again, whether she wanted to marry Clem. She did not. Did she want to marry someone else at Suabi, Leo perhaps? She did not. She just wanted to return to her family. Then, Waiyo, argued, she should go at once. People at Suabi should raise the money for her airfare to Mount Hagen. A flight to Tari would have been much cheaper but, for a young woman on her own, travelling by bus from Tari to Mount Hagen was dangerous. No one disputed that. Tuppy should go. But, Waiyo added, Epe should stay. His presence at Suabi was to be the guarantee that repayment to Clem would be eventually forthcoming. Thus it was agreed. People at the hearing contributed money. Leo himself was the first with a substantial offer—PGK400 of the PGK700 that was needed.

Waiyo had acted with haste, for good reason. She knew, but many did not, that Leo and Tuppy had had sexual relations. She did not want that behaviour to continue. And she certainly did not want Clem to know. She called the court case before the information could spread. She talked over the top of the senior males who adjudicated the case. And she achieved what she wanted. By suppressing the full story she protected her son's interests and put in place the departure of a young woman who now, after denying Clem but having sex with Leo, was a potential threat to the latter. Epe remained and, with help from Suabi residents, found employment at Juha.

While there was certainly appeal to expectations of 'proper' practice in Waiyo's argument, it was of little direct relevance to her motivations. She was deeply concerned with concealing aspects of the present. But where the past is the focus, the discussion at one event may lay the ground for future cases. The exchange of money for children that opened this chapter was a case in point. Two Yawuasoso households hosted the day's activities. A man from one of those households shot the pig, a man from the other received most of the money. On that day these two households worked together. But they heard and, indeed, contributed to talk of failed marital exchange. They, and their guests, were reminded of another marriage that had not been reciprocated. This was the union of Wuagodua's parents, a marriage between classificatory brother and sister since both were Yawuasoso. The marriage was against convention; to Kubo it was incestuous. Such a marriage cannot be reciprocated—the 'sister' lost cannot be replaced—without compounding the transgression. Now, however, after a lapse of at least 35 years, there was talk that payment was due to the woman's brothers. Their claim gained more force because, in 2013, in preparation for clan vetting, the senior men of Yawuasoso chose to upgrade three subclans to the rank of major clans. Under this restructure the marriage was no longer between people of the same clan, it was no longer incestuous and it was no longer advisable to refrain from controversy by ignoring past misdemeanours. Many who spoke to us expected that this matter would soon be aired in court, with the two households who had collaborated in hosting the exchange for Wuagodua's children now on opposing sides.

Affiliations

Among Kubo, the Papua New Guinea Liquefied Natural Gas (PNG LNG) Project has prompted a reconfiguring of social categories. This, in turn, as seen in the previous example, has provided opportunities to resolve long-standing tensions concerning marriages judged to have been incestuous and, hence, at that time, unable to be balanced by sister exchange. Here, then, people at Suabi were navigating the past through reimagining relationships with and between collectivities, not just between individuals. Such reimaginings could be expressed with respect to both marriages and the use of land. In both cases, however, there was some awkwardness and uncertainty about how to reshape the past in ways that were appropriate to the present.

At Suabi, in the years 2011 to 2014, issues of rights to land featured in many conversations. These concerns were stimulated by the heightened focus on the future incorporation of Land Groups. It was early days, however, and few people knew precisely what land was being 'claimed' by others. In only one case, and it had not yet come to court by mid-2014, was a pending dispute aired publicly. For the first time, and in contrast to the 1999 dispute at Mome Hafi (Chapter 2), the claim was asserted solely on the basis of past genealogical connections. It was more than 35 years since any of the claimants, or their forebears, had used the land.

In 1986–87, at Gwaimasi, we learned that a stretch of land on the east side of the Strickland River just across from the village had, formerly, been occupied by a branch of Kesomo clan. We were told that there were no living members of this branch, and the land was regularly used for hunting, gardening and sago processing by members of Gomososo clan. By 2014, however, members of another branch of Kesomo whose land was some distance away at the headwaters of Toio Stream laid claim to that land, naming the 'landowners' as Ia Hafi Kesomo—Ia Hafi marks the junction of Ia Creek with the Strickland—and asserting that a recently deceased Gwaimasi woman, Fafobia, had been the last representative of this group. Ia Hafi Kesomo was not mentioned in the booklet *The Origin of Kesomo* compiled by Henick Taprin in 2007–08 but, under the name 'Iyohafitie', was listed and accepted as a subclan of Kesomo at the 2013 clan vetting meeting. From 1986 to 1999 we were never in doubt about Fafobia's membership within Gomososo. It was expected by people at Suabi that the grievance developing between Kesomo and Gomososo would eventually need to be settled in a local court hearing. But people were aware that, on one side, the protagonists would assert their rights by reference to genealogical connections (perhaps fabricated) while on the other side, the protagonists would assert their rights by reference to their use of the land for more than three decades. Non-Kesomo people who discussed this case with us considered that rights through ongoing use had more validity than distant genealogical links and, while acknowledging uncertainty about the eventual outcome, suggested that 'government' would see it this way too. The eventual outcome of this case may well set the scene for airing analogous differences in the future.

In 2012, Alex assisted in the preparation of a list of members for an Incorporated Land Group (ILG) that people planned to register for the clan Osomei. On that list, Alex took Bugawo as his father's name. His earliest memories were of these associations. But now Levai spoke to him, reminding him of his past. Alex's biological father had died soon after Alex was born, and his mother had then married Bugawo. Levai was a senior Busuo man and, he said, Alex's birth affiliations were to another branch of Busuo with land to the west of Suabi, far from Osomei land. Indeed, Alex's branch of Busuo was the originating branch and Alex was the only surviving adult male. Levai argued that Alex should lay claim to this land by reassuming his biological father's name, Gubia, and devising his own Busuo ILG list. From that time onward, Alex presented himself as Gubia's son and in 2014 drew up a separate ILG list to represent the interests of his branch of Busuo.

Past genealogical connections that, sometimes for decades, had lost salience often became important tools in either asserting rights to ILG membership or rationalising inclusion within an ILG list. For example, in the mid-1960s, at the time when the government station at Nomad was well established, an unmarried Yawuasoso man named Moiyo attended a feast hosted by a Samo community. In the course of his visit, he had an affair with a young woman who became pregnant and gave birth to a son. Moiyo departed. The young woman married and her son Hobesa grew up as Samo. In 2014, Yawuasoso people were drawing up ILG lists and needed names that they felt could legitimately be taken to represent newly devised subclans to be associated with land in the Juha area. Yameka told the story of his father's past adventures, so Hobesa was invited to represent and prepare his own ILG list for one of those subclans.

In the 1940s, as a child, Sinage's father and another boy were accused of sorcery. Their affiliations at that time were with the Febi clan, Gumitie. The boys were banished instead of being killed. Sinage's father was invited to stay with Strickland River Headubi people. They gave him land that had formerly been occupied by a clan named Bigiti but that no longer had living members; we were told that they had been raided and killed by head-hunters who came up the Strickland River from the vicinity of Lake Murray. Sinage's father did not return to his natal land. When he died his widow remarried a Headubi man and Sinage grew up identifying with this *oobi*. He married a Gomososo woman and lived with her brothers—one of whom married his Headubi classificatory half-sister—at Sesanabi and then Gwaimasi. In the 1980s and 1990s, he identified as both Headubi and Bigiti, never as Gumitie. By 2013, however, his children—the oldest daughter was then about 33 and the oldest living son was 26—identified variously as Headubi, Bigiti or Gumitie. As Wage told us, it depended on where he was and who he was with. When they visited Suabi, they now chose to stay with Gumitie 'brothers' or 'sisters'. In earlier years they had not done this. In 2014, Wage and one of his brothers each drew up an ILG list supposedly for two distinct groups within Bidibi, a subclan of Gumitie. At the Siabi clan vetting meeting, however, Wage and his immediate kin did not receive the support they expected from those with whom they now asserted affiliation. While the name Bidibi appeared on the list of recognised clans, as a subclan within both Gumitie and Hwotie, the named representatives did not include Wage, his father or his brother.

In addition, people sometimes drew on mythological connections to establish rights or to justify amalgamation. For example, Headubi people who live south of Suabi had been invited to affiliate with Andibi, in the Suabi group. But Headubi people to the north, in Febi territory, were seeking to fill out their own lists and suggested it would be more appropriate to align with them. Their argument prevailed. The origin stories of both groups were similar. They had descended from 'dog'. The southern group told of a time in the past when, near the Strickland River, which they took to be the origin place of all Headubi, dogs were fighting over a bitch in heat. One dog, chased by the others, fled into a cave. It emerged on a small island in the Damami River, 25 km to the southeast. That dog's descendants are the Damami River Headubi whose logo depicts this tale. Another dog, perhaps fleeing the same fight, emerged from a hole near Tobi, where the Febi Headubi now live. Martin told us of seeing that hole not long ago.

Again, Osumitie had been listed as legitimate landowners in the Juha area—recorded as Febi rather than Kubo—but there were no living members of that clan. People from neighbouring clans, outside the domain of primary landowners, feared that they would be excluded from sharing in the benefits to flow from the PNG LNG Project. At the Siabi clan vetting meeting, Department of Petroleum and Energy officers suggested that the land of Osumitie was without people, so should not be included in the distribution of benefits. 'We are here,' Martin called. 'We are fire clan.' And he named six groups that he asserted were subclans within the major clan Osumitie, groups that were united in being the first people to find fire and use it for cooking, and were additionally united in sharing a 'special place': a *toi sa*, a 'forbidden place' where the spirits of the dead reside. Martin was stretching a point. At best, only four of these named groups had rights to either the origin story as he told it or to the *toi sa*. At worst, it was only two, which in our previous understanding had together constituted a stand-alone *oobi* to which Martin belonged. One of the two groups that were clearly outsiders—invitees—has close agnatic connections with one of the insider groups. The other, Dobiti, is an amalgam of seven named groups—one of which retains the name Dobiti—that, 20 years earlier, we knew as four stand-alone *oobi*. These groups tell how fire reached them from the north and assert that they share a number of 'special places': one associated with a root that kills fish and can kill people, one where people can ask the spirits to send sickness to someone that is doing them harm, one where an old woman hid fire before a palm cockatoo stole it and spread it to everyone, and a fourth where, if a person visits and sleeps, a waterfall spirit may foretell the future. Dinosi told us that he himself went there, and dreamed that he would never be a good hunter. So now he doesn't try to hunt pigs—he admitted on a different occasion that he was scared to do so—but, instead, hunts smaller animals: birds and fish. Primary responsibility for the care of these places usually belongs to just one of the seven groups but all, more than people from other groups, have access rights to them.

Further, in at least one case, options were kept thoroughly open by blending the mythological with overt commercial possibilities. Men from Yudo clan, well to the south of Suabi, tell us that long ago two brothers went diving in a secret place and saw a gudgeon of a type named *sa*. One of the brothers shot at the fish, but missed and his arrow stuck in a log. He dived under the log, came up on the other side and found that he was now in a large lake. He was frightened, dived back and returned to the secret place. But he could not get out of the water because he had turned to stone. He called to his brother who came running, saw what had happened and wanted to lift his stone-brother out of the water and carry him home. The stone-brother said to leave him in the water and bring food. And that is what happened. To this day the stone-brother is still in the water at that secret place. When the stream floods the stone-brother becomes light and floats to the surface of the hole in which he lives. Around him float leaves of many different kinds of food plants: bananas, sago, yams and taro. The leaves are from the food that his brother brought to him. But when the water level drops the stone-brother gets heavy and sinks out of sight.

In 2014, we were visited by two Yudo men who reported their own recent experiences—experiences that, in part, recapitulated the mythological tale of the stone-brother. They went diving for fish at the secret place. One of them pursued a *sa* that went under a rock. It seemed to be hidden by bark. The man put his hand under the rock and started to pull the bark out. At first it was very heavy but as it came up to the surface of the water it became much lighter. It looked like a piece of wood, but closer inspection revealed it to be a 'stone' that was shaped like a man, with distorted trunk and head, foreshortened limbs, no feet or hands, and an indentation in its belly where its navel would be. The two men kept the stone and took it to Testabi where they lived. On the night that they did this there was much lightning and thunder and, in the storm, the 'stone' called like a frog. Children were frightened and could not sleep.

The Yudo men were very enterprising. We did not see the stone-man but we did see photographs of it. The 'body' was as large as that of a man. But it was clearly not heavy. It was decorated with brown lines that radiated out from the 'navel' and criss-crossed the appendages; marks that bore some resemblance to the body painting on Kubo dancers or Samo initiates (Shaw 1990: see plates following p. 114). The lake in the origin story was said to be Lake Yubi far away in Yawuasoso land and the re-emergence of this story was intended to link Yudo—'stone clan'—with Juha landowners and, hence, legitimise a claim to future benefits from the production of gas. Some Yawuasoso people said they had never before heard this story. But the Yudo men had additional commercial interests. At their home village they were charging an admission fee before letting people see the stone-man. And they were offering it for sale to white people who heard the story, recognised its significance, and might make a satisfactory offer of money.

Distinctions

The entrepreneurial spirit of the Yudo men—in mobilising stories and symbols drawn from the past to not only negotiate affiliations with distant clans but also render these symbols themselves negotiable for money—hints at an emerging objectification of the past that renders its expressions both ownable and alienable. By mid-2014, though people at Suabi had not registered any ILGs, they had drawn up many plans and named many potential groups. They were, in this sense, corporatising social identity. And, like knowledgeable corporate citizens, they sought to enhance potential visibility by creating logos that would serve as iconic representations of those newly formed groups. They were branding identity (Klein 2000; Comaroff and Comaroff 2009). The design of these logos provided yet another domain in which people at Suabi selectively drew on the past to facilitate access to a desired future.

People here were familiar with logos. The ECPNG mission had been based at Suabi for 25 years, planes from various airlines touched down at the Suabi airstrip and, from time to time, oil and gas companies established base camps or undertook exploration nearby. Nearly all these groups self-brand with a logo that captures some essence of their aims and purpose. Documents produced by the nation state, the provinces and government departments are commonly embellished with a logo (Fig. 6.3a).

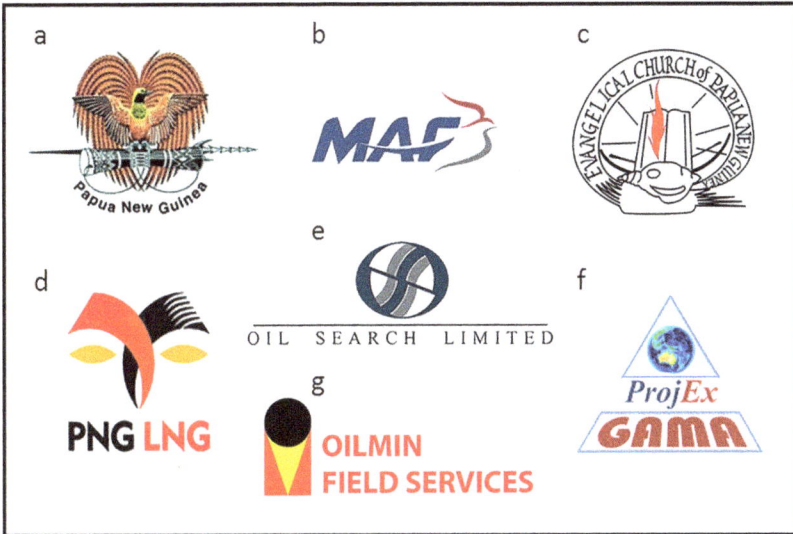

Figure 6.3: Logos: (a) Government of Papua New Guinea; (b) Mission Aviation Fellowship; (c) Evangelical Church of Papua New Guinea; (d) Papua New Guinea Liquefied Natural Gas Project; (e) Oil Search Ltd; (f) Gama ProjEx; (g) Oilmin Field Services.

Sources: Public documents and websites for the named organisations.

At a smaller scale, a number of organisations established by Papua New Guinean nationals intrude in various ways on the lives of people living at Suabi. The Gigira Development Corporation, established by Huli-speakers in 1991, provides a variety of services—labour, security, cartage—to companies associated with the PNG LNG Project and, from time to time, has employed men from Suabi. It is named for Mount Gigira, at the headwaters of Baiya River. A myth and ancient prophesy tell that, within the mountain, there is a perpetual fire and that the time will come when a man with 'orange legs' will try to take the fire from those who rightfully own it (Cuthbert 2009; Bashir 2010c).[3] The corporation logo features the mountain (Fig. 6.4a). Juha Resource Development Corporation Ltd, established in the mid-2000s with the intention of accessing business and infrastructure grants associated with the PNG LNG Project, identifies as 'the umbrella company representing seven [Hela] tribal groups of the Juha LNG Project'.[4] The company logo features Mount Gigira, the founding

---

3   In other versions of the Gigira myth, the man who comes from the outside is described as 'white-legged' or red-legged (Garnaut 2015; Michael Main, personal communication, 31 October 2015).

4   Viewed 28 October 2015 at: www.facebook.com/juhalng/timeline.

ancestor Hela, and gas flaring from both Hela's head and representations of well heads (Fig. 6.4b). By at least mid-2008, Gas Resource Juha Limited had been established by one Febi man under the patronage of several Huli men. The company mimicked the aims and intentions of Gigira Development Corporation and Juha Resource Development Corporation Ltd. For a brief period, in the later 2000s, it provided employment for a few Suabi men but, by 2010, it was no longer active. The logo is shown as Figure 6.4c.

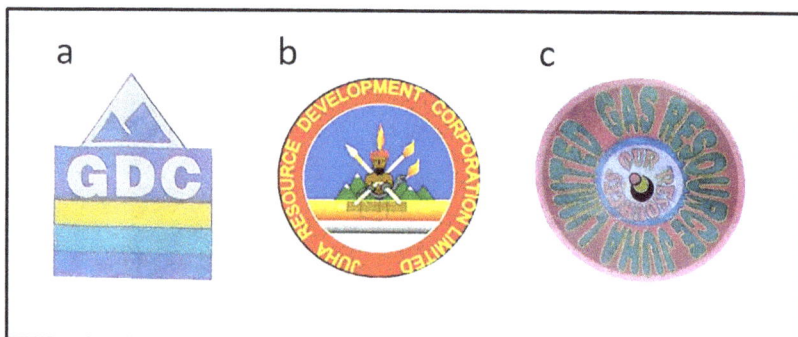

Figure 6.4: Logos: (a) Gigira Development Corporation; (b) Juha Resource Development Corporation Ltd; (c) Gas Resource Juha Limited.
Sources: Public documents and websites for the named organisations.

By 2011, then, people living at Suabi appreciated the importance of logos with respect to enhancing visibility and were familiar with a range of logos that might serve as models for those that they themselves created. They drew logos, or planned to draw logos, to represent ILGs, Business Groups, women's groups, schools and church groups. With the exception of the single Church logo that we saw—an open Bible and a cross (Fig. 6.5a)—these depicted local resources, subsistence activities or mythological origin stories. For example, two logos drawn for the Suabi Community School feature, centrally, a length of partially rolled bark. Towards each end there is an arm band—a decorated *dobe* of the kind that men wear when they dance—and, held in place by one of these, a cassowary bone dagger. Two stone axes hang over the roll of bark, and emerging from it are two hunting arrows, a bow, a hand of bananas and a sprouting yam (Fig. 6.5b).

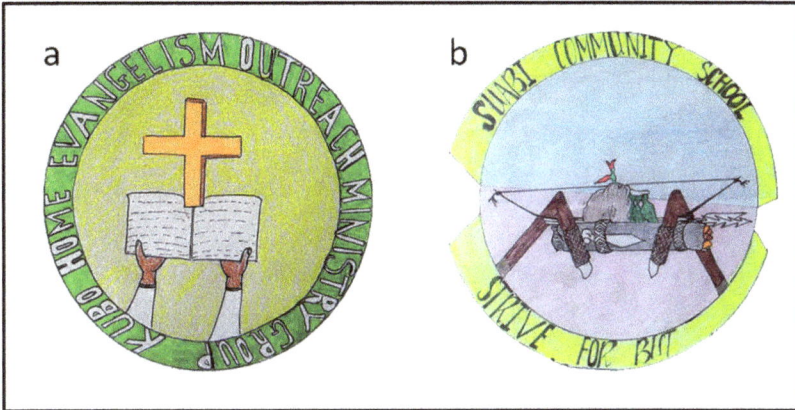

Figure 6.5: Logos: (a) Kubo Home Evangelism Outreach Ministry
Group; (b) Suabi Community School.

Sources: Originals provided to authors for preparation as electronic images. Reproduced with permission.

In 2013–14, a number of young women at Suabi felt frustrated that they were poorly represented in the plans to form Business Groups that men were drawing up. They were seldom represented in either the intended aims or the organising committees of those groups. Some women chose to initiate a similar process without input from men. Four such groups were named Yanowosu Young Girls, Tobi C.B.C. Women Group,[5] Juha Mama Group and, from a different Kubo community, Kusobi Women's Business Group. The aims of these groups were modest. They planned to seek funds for the purchase of sewing machines, cloth, needles and cotton that would allow them to establish businesses through which they would sell clothes at local markets. One group planned, in addition, to develop a commercial agricultural venture through which rice would be sold to local buyers. This group—Kusobi Women's Business Group—stated that it had been established 'due to major LNG projects that is coming up in area'.

Yanowosu Young Girls was initiated by five women, married and unmarried, twenty to mid-thirties in age, affiliated with three different clans—Yawuasoso, Nomo and, by marriage, Osumitie. Several had completed Grade 9 at school. The name of the group combines the initial letters of the three clan names ('Ya', 'No' and 'Osu'). The logo features

---

5   C.B.C. is the Christian Brethren Church which, since the mid or late 1980s, has been established at several Febi communities.

a variety of palm characteristic of higher altitudes (Yawuasoso) and fire, which is said to have originated with Osumitie but is also mythologically important to Nomo (Fig. 6.6a).

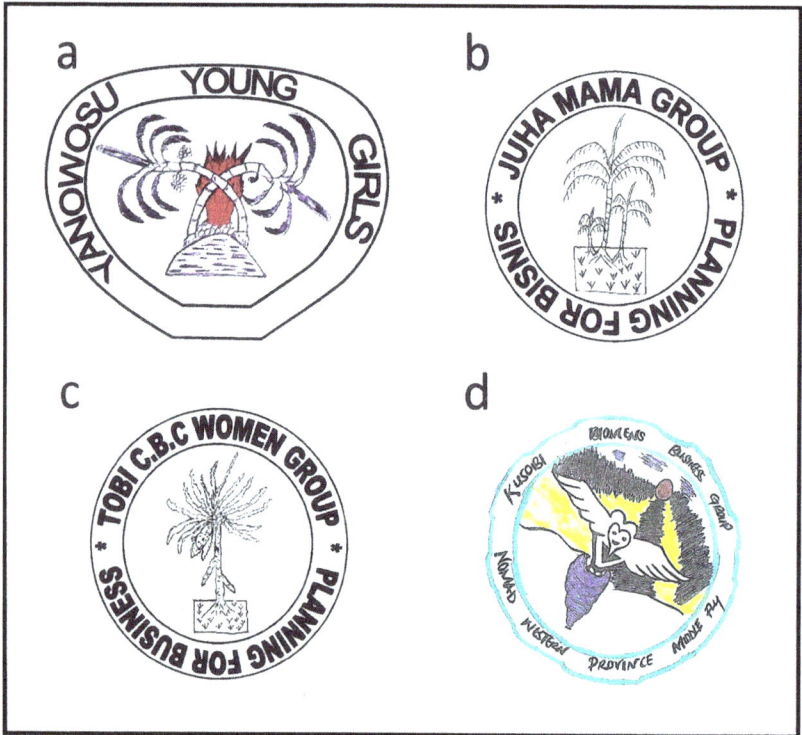

Figure 6.6: Logos: (a) Yanowosu Young Girls; (b) Juha Mama Group; (c) Tobi C.B.C. Women Group; (d) Kusobi Women's Business Group.

Sources: Originals provided to authors for preparation as electronic images. Reproduced with permission.

Note: The outlines and words on (a), (b) and (c) were completed by the authors according to the instructions of members of each of those groups.

Membership of both the Juha Mama Group and Tobi C.B.C. Women Group was restricted to women born to, or married into, particular broadly defined, higher altitude Febi clans and the key symbols used in their logos were, respectively, a variety of black palm and a variety of fruit pandanus said to be 'held by' members of those clans (Figs 6.6b and 6.6c). The fruit pandanus depicted had originated at lower altitudes but when mythologically transplanted to the mountains the fruit that it produced changed colour from day to day—sometimes red, sometimes green and sometimes yellow. There were spirit connotations and some older women were unsure that it was a wise choice for a logo; this was a *masalai* pandanus, they warned, and thus dangerous.

The Kusobi Women's Business Group was formed in 2013 by women living at Kusobi village and affiliated by birth or marriage with four Kubo clans. The logo for this group featured an eagle and a stone that are implicated in an origin myth of the primary clan—Seaso—with which the women concerned were affiliated (Fig. 6.6d). These symbols were used also by male members of the same clan when preparing an ILG list.

On 23 February 2008, men from Suabi, representing at least seven clans, registered ABBA Co-operation Limited. In the course of the next two years, with outside assistance, they submitted an elaborate application for PGK5,000,000 as a 'seeding grant' with the aim of developing commercial enterprises in, among other areas, agriculture and livestock, aquaculture and the supply of educational materials. The company was awarded PGK20,000. This money was used to support several men who relocated to Port Moresby. None of the commercial enterprises came to fruition. ABBA Co-operation Limited was extant, but not functioning, in the years 2011–14. It did not have a logo.

The work entailed in drawing up membership lists for ILGs was, for the most part, done by men. It was men too who chose the logo that would represent these groups. Elements from myths—gendered stones, forbidden places, mythologically significant plants and animals and landscape features (rivers and mountains)—were prioritised in the logos that we saw or were told about. An ancestral crocodile for the subclan Dobasoso (Fig. 6.7a), a bat implicated in the enmity of two clans for Osomei (Fig. 6.7b), the dog that emerged from a cave on an island in the Damami River for Headubi (Fig. 6.7c) and a varanid lizard (Wabogosai) and palm (Kowa) for Mithy (Fig. 6.7d). In the last case, the name of the lizard was taken as the name of the intended ILG. It is a small lizard that shows its forked tongue if it 'is happy to see you'. The upper half of the logo features the small black palm that splits into male and female trunks. The male trunk is identified with Wabogosai and, in the background of the palm, there are two mountains where the spirits of dead Mithy people reside, where Kowa grows and Wabogosai lives.

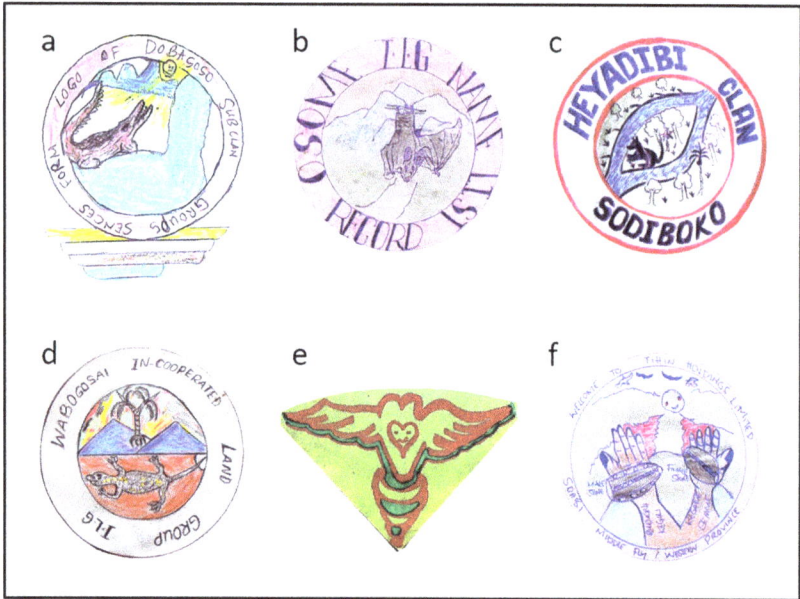

Figure 6.7: Logos: (a) Dobasoso subclan; (b) Osomei ILG; (c) Heyadibi clan; (d) Wabogosai ILG; (e) Seaso clan; (f) Tihin Holdings Ltd.

Sources: Originals provided to authors for preparation as electronic images. Reproduced with permission.

Tihin Holdings Ltd was a Business Group that men of Seaso clan planned to register. Their elaborate documentation took the form of an application for funds to build a classroom and two houses for teachers at Kusobi, a guest house and nature lodge at Suabi, and the wherewithal to initiate a commercial venture in 'handy craft' and artefacts. The logo that accompanied this proposal featured two named stones, one representing a son and father, the other representing a daughter and mother and both implicated in the origin of the clan (Fig. 6.7f). These men had drawn another logo—an eagle and stone—that they intended to use when registering a clan-based ILG (Fig. 6.7e; see also Fig. 6.6d).

The Osomei clan logo shown as Figure 6.7b was drawn in 2012. In 2014 it was revised. Alex, the community health worker, had purchased a laptop computer and, from time to time, was granted access to the internet by the medical officer at the Company camp. He used these facilities when redesigning the logo. And he told the following mythological story:

Flying foxes lived in a cave in limestone country. In the hours before dawn, returning to their sleeping cave, they cross a saddle in a ridge. Here men would build platforms and, using switches, knock down the bats as they flew across the saddle. One day, when he and his family wanted to eat flying fox, Sada built a platform intending to hunt before dawn the next morning. He went home to wait for the right time. In the meantime, however, another man, Kodo, came along, saw the platform and decided to use it. When Sada arrived he saw that someone had usurped his platform. He called out, asking who was there. He got no reply. He was angry and shot Kodo, killing him. That was the start of fighting between those clans.[6]

Alex's redrawn logo—it was not finished when we saw it—was circular. Framing the circle were the words 'Fighting for Flying Fox'. In the centre, Alex was inserting part of a painting that showed 'dark skinned warriors' carrying a man trussed to a pole in the way that Kubo people carry a dead pig. There could be little doubt that the victorious warriors had cannibalism in mind. Alex had found the painting on Wikipedia, under the caption 'Cannibalism'. The new logo did not include an image of a bat (compare with Fig. 6.7b).[7]

Zavia had not yet produced a logo for the subclan of which he was to be key representative. But he knew what he wanted though he said that he himself was no artist and would find someone else to draw it to his description. It was a bit complicated, he told us. He wanted to show a lake, a big tree, a boy on one side and a girl on the other, a bird on top, a frog down below and perhaps an old man. These, he said, would tell the origin story of the subclan Tafenden within the clan Mogotie. The story was as follows:

There was a brother and a sister and an old man. The children used small bows to hunt grasshoppers and geckos, which they then cooked. The old man told them not to do this, and pissed on the fire to stop them. The children went out again. When they returned with their catch and saw the fire was out, they started to cry. The old man then began to make fire again, in the 'traditional' way. As he rubbed stick and string together, there was an earthquake. The ground cracked and water started

---

6 The enmity was between the Kubo clan Osomei and a Bedamuni clan. In the course of a revenge attack some Osomei people fled underground and emerged at a cave on Gomososo land near the Strickland River. They and their descendants became known as Woson, though, through long-term association with Gomososo, they were accepted as 'brothers' to members of this clan. Another underground passage from Doitafa cave also connected the lands of Gomososo and Dumiti clans until a time when it was closed by a tree fall that a bird failed to prevent (Dwyer and Minnegal 2007: 551; Minnegal and Dwyer 2011a: 327, n. 2).

7 The bat shown in Fig. 6.7b is a cave-dwelling microchiropteran within either the family Hipposideridae or the family Rhinolophidae. The relevant myth, however, is about a species of megachiropteran—a cave-dwelling bat of the genus *Dobsonia* in the family Pteropodidae.

to come up. The boy and the girl were swept away. They swam and swam but the water kept coming and kept getting wider. It was becoming a lake. At last the children came to a big tree. They climbed into the branches. The boy climbed to the top. The girl stayed in lower branches just above the water. Now the lake was very wide and they could not see land. They were frightened. Then the old man—their grandfather—came out of the water. He told them not to be frightened. They should come into the water with him because there they would find many good things: sago and bananas and all sorts of other plants that provided food for people. But the children would not go with the old man. So he returned to the water by himself. He collected leaves from all kinds of plants—from bananas, sago, fruit pandanus and others. He brought these to show to the children so that they would understand that he was not tricking them. Again, he asked them to come into the water with him. But the children still said 'no'. Then the boy turned into a bird (Kuauwe) and the girl became a frog (Tiando).

The lake that comes into being in the course of this story is Lake Yubi on the land of Yawuasoso. Kuauwe is the butcher bird whose dawn call signals that dances that have continued through the night may now come to an end. And the old man, Zavia told us, was named Dibodoin but his own origin place was not in the mountains where the lake was formed.

Dibodoin originally lived down below, where the swamps are, not far from the Strickland River, on the land that is occupied by a different branch of Yawuasoso. It was another man—Kofainsa—who originally lived in the mountains. But these men changed places, each taking resources from one area and distributing them to the other. Dibodoin travelled from the swamps to the mountains. As he travelled he discarded bits of his clothing and some of his possessions. His nose-plug was tossed aside. It grew into a very big bamboo. His clothes, which were made from bark, were also thrown aside. They became different kinds of trees. And some sago flour that he carried was poured onto the ground near the place where Lake Yubi formed. It grew into a kind of sago palm. Food resources that you would otherwise expect to find only in the lowlands were carried into the mountains, and distributed there, by Dibodoin.[8]

---

8   Kubo and Febi people continue to move plants from one place to another. At Suabi we were shown plants that had been transplanted to the village from the high mountains beyond the altitudinal zone where people lived and from Kiunga, Highland and coastal towns. Selection of plants for transplanting was motivated by both utilitarian and aesthetic reasons.

For Zavia, all these people and events, either explicitly or by the visual prompts provided, would be expressed by his intended logo. Some were given wider expression. Zavia had assigned the names of the two mythological figures, Dibodoin and Kofainsa, to two boys when he drew up his list of ILG members. He had assigned the latter to a boy from the Strickland River branch of Yawuasoso, and thereby—since the original Kofainsa had migrated from mountains to lowlands—tied that boy more strongly to Zavia's own mountain branch of the clan.

In 2014, Zavia was about twenty-eight years old. He had grown up through years when origin stories were declining in relevance, when his adult kin were moving towards new ways of understanding the world, when their focus was often with Christianity, Bible stories, sin and its consequences, and the possible arrival of a petroleum or logging company that would enhance material well-being. It was only recently that he and his age mates—the young men and women who could write, read to a limited extent and grapple with the requirements of bureaucracy—came to appreciate the importance of the past in asserting rights to the future. They turned to their parents to hear and learn the stories. As one young man told us, he had spent so many years away from the village at high school that he did not know these stories. Did we know the stories of his clan, he asked. We ourselves had first come to the land of Kubo people around, or before, the time that most of the more formally educated people had been born.

The stories that Zavia planned to represent in his logo were, however, those of all Febi and Kubo Yawuasoso. Zavia was appropriating those stories to a named subsection of a section of that greater clan. In doing so, he was laying claim to symbols that actually belonged to many more people. Other subclans in Yawuasoso, he told us, would need to devise different logos. Choosing a logo was not easy and had potential consequences for other people. Members of Nomo clan, for example, had not yet settled on a design for a logo to represent a company they planned to establish. They wanted to find a 'common element'—an animal, perhaps, or their *toi sa* (forbidden place) that spoke to their uniqueness. They chose not to use their *toi sa* because, they argued, some Highlanders assert that their own ancestors came from that place and so they might have a claim to the company. They thought of using 'fire', since they too are a 'fire clan', but Martin had already taken 'fire' to represent his own clan. And they thought of using 'pig'—one Nomo story tells of a pig entering a cave and emerging in the land of Bedamuni people—but, again, learned that they had been pre-empted by another group. Similarly, the decision by Damami River Headubi to use 'dog' in their logo (Fig. 6.7c) had implications for others in the constellation of the six clans identified as Headubi, all with origins traced to dogs, holding non-contiguous areas of land within Kubo, Febi and Konai territories.

As Zavia's account and those of Nomo and Headubi people illustrate, origin stories are not discrete. They are not specific to bounded sets of people. Like clans' lands as they were in the past, these stories flowed into one another; they grew, diverged and borrowed from each another. They separated some people and united others, but the similarities that emerged and the differences created through the telling of these stories were always in flux. As Frederick Barth (1987) argued, cosmologies are not fixed, they are made. 'There were never any land boundaries in place to separate the four Kesomo sons from each other' Henick Taprin wrote in *The Origin of Kesomo* (2007/8: 6). Now, however, as with land, people are beginning to creatively define the limits of the cosmological; they are bounding it, declaring selected stories or selected constructions of stories to be properly and exclusively their own.[9]

Distributions

In late February 2014, representatives of Talisman—the company then using the exploration camp at Suabi—arrived by helicopter to make what they considered the 'final payment' for rent and for environmental damage caused by their recent activities. The latter arose from felling trees and loss of gardens in and near the area where the camp was established. In at least one case, the garden in question had been prepared and planted in the hope of harvesting money, rather than food plants, if Company did in fact arrive.

The payment for rent—PGK6,400—was received by Ifan, one of the senior men in the seven clans that were considered by local people as the deserving recipients of the money. These clans were Andibi, Baiyameti,

---

9   Lihir Islanders take the rock Ailaya as both an origin place and a final resting place for human and spirit forms (Bainton et al. 2012: 33). The appropriation of the Aiyala by several Liberian groups (e.g. on a logo of a local cultural heritage organisation, and on the front of a new business centre) has generated concern over possible implications for access to wealth from the Lihir gold mine. Holly Wardlow (2004: 54) records use of the mythologically important Mount Kare python as the logo of a Huli-owned trade store.

Domiti, Gobogometi, Osomei, Sisiti and Wamiti. It was people from these groups who, in the early 1980s, came together to form Suabi. Only three of them qualify as genuine landowners of parts of the land on which the airstrip and its approaches were made, and probably only one—with a single living representative—as a genuine landowner of the area utilised by the exploration camp.

The payment for environmental damage included PGK2,945 for felled trees and PGK1,370 for loss of gardens. The money was received by Kamuna, the local Mission Aviation Fellowship agent, whose classificatory mother had lost a small garden as a result of Company activities. The package containing the money named three clans as intended recipients: Domiti, Osomei and Sisiti. But members of two other clans, Wamiti and Djagososo, had also lost gardens; they had been 'forgotten' we were told.

Ifan and Kamuna called a meeting at which people discussed ways in which the money should be distributed. This was no simple matter. Some argued that the compensation money should be shared equally to all clans represented in the Suabi community. It was, after all, merely fortuitous that, when Company arrived, particular people had established gardens on that land; had Company come earlier or later it would have been others whose gardens had been lost. This suggestion was not adopted. Different people had different sorts of claims, and drew on historical connections to support those claims. While it was accepted that the primary claimants to rent money were members of the seven clans listed above—it was their forebears who had established Suabi and, 30 years earlier, contributed most work to making the airstrip—two of those named groups are, in other contexts, treated as subclans of a greater entity, Nomo. It was felt, therefore, that some money should flow to members of other groupings within Nomo. And, again, though Kamuna himself had not lost trees or gardens he had assumed responsibility for keeping the money secure, for calling the meeting, and for providing the venue at which it was held. For these contributions he received PGK100. Present circumstances and remembered pasts came together as people decided who should and who should not receive a share of the money.

The meeting was held that night, but not everyone had attended. The task, the next morning, was to deliver allocated shares to people or groups who had not been present. Ifan distributed rent money. But he was anxious. He wanted to get the job done quickly so that those who were to receive, but had not come to the meeting, would soon know that they had been included. He was concerned too that some who did receive might feel they merited more than they got, and that some who did not receive might be offended. He did not like handling money of this sort, he told us. He was in favour of giving equal shares to everyone; then he would not have his current responsibility. If he delivered money to people who were dissatisfied then they could be cross with him. They could take him to court though this would not resolve the issue. Even if they won they would remain cross. They could wish him harm. He could be ensorcelled. In the present world of money, paper and writing, he told us, there were new forms of sorcery to worry about. You might be walking alone to a garden or bush house, perhaps, or even in the village. Someone comes towards you, crosses paths with you. They stop and talk. But their talk is *gaiman*—it is 'grease talk', it is false. They are distracting you while others, who you do not see, creep up behind you and 'kill' you. You are dead but you are not yet finished. You go home and, only later, get ill and finally die. It was not the same in the past, Ifan complained. Then, sorcerers worked alone. You would see the sorcerer face-to-face—just one—and there was a chance you could elude him. Now they roam in groups, and you do not recognise them for what they are. You no longer have a chance. And, perhaps, the sorcerers themselves do not know what they are doing. They may be young men or young women who have found paper on the ground. They have picked up the paper, read the words on it, and their minds are taken over. They become a new kind of sorcerer. When money comes from outside, when it is necessary to share it responsibly and fairly, there is always a concern with sorcery, with new forms of sorcery, adaptations of past forms that have emerged in, and been accommodated to, new contexts.

Ifan had favoured an earlier style of distribution, where all—perhaps not each person but certainly each household, or *oobi*—would share equally in the benefits derived from the land. But a new logic was intruding now, with money at stake. Evaluating differences in ownership and entitlement was an increasing concern at Suabi. But while 'fairness' might now be read differently, past understandings continue to permeate both its importance and the anticipated consequences for failing to meet these new expectations.

* * *

In resolving disputes, accommodating marriage arrangements and concerns about sorcery to new ways of living, drawing up ILG lists that reshape social arrangements, devising logos that will stand for newly conceived groups of people, and finding ways in which money may be distributed more fairly across the community, Kubo and Febi people draw selectively from their remembered pasts, both secular and sacred—often foregrounding some details while suppressing others—to fit their lives to an ever-changing present. In striving to achieve these ends—in 'the forging of new futures'—they engage in creative work whereby present conditions may be transcended and histories rendered effective (Bell 2016: 22). To people at Suabi, however, the present is deeply imbued with desires that are oriented towards a future, a future in which 'development' comes, in which they are no longer 'remote' and forgotten, and in which wealth that is perceived as 'rightfully' theirs is given material expression either in the form of extractable resources—gas, oil, minerals, timber—on their own lands or rights to the benefits expected from such resources on the lands of their immediate neighbours. In the final analysis, Kubo and Febi people draw on their remembered history of engagement with land and with each other as part of a strategy for navigating the future.

7. The Giving Environment

By late April 2014, Talisman had finally decided that the exploration camp at Suabi would be 'demobbed'. Stoves, refrigerators, toilets, showers, marquees, generators, pump and water purifying system, a small forklift, tools, fuel drums, satellite dishes, computers, printers and so forth would be gradually packed into containers and moved by helicopter to either the Talisman base at Yavo on the Strickland River or a new exploration site more than 45 km to the northwest. First, however, it was necessary to carry out an audit of everything that remained. Sean, who was then the resident camp manager, had been assigned this responsibility. In one container he found a large amount of rice, old stock that was now infested with weevils. He checked with the camp medic who assured him that if the rice was well cooked it would be safe to eat. So he gave it away: eight bales—each with 20 kg of rice—to the Evangelical Church of Papua New Guinea (ECPNG) Mission and another eight to the Seventh-day Adventist Mission. He included cartons of chicken noodles that were also past their recommended 'use-by' date. He told us, however, that as the old rice was being handed out those who were assisting managed to filch quite a lot of good-quality rice. They were quick; he never saw who it was that took the rice. A week later, auditing a different container, Sean was disappointed to discover that many files and hacksaws were missing. Local men, employed as labourers and security guards, had been helping themselves to Company belongings.

People at Suabi had high hopes for future benefits from the Papua New Guinea Liquefied Natural Gas (PNG LNG) Project, in the form of royalty payments and business development grants. They understood that these would be provided to those who—through incorporation as legal actors—rendered themselves visible to the structures that governed distribution of benefits. But it was the presence of 'Company' on their land, and undertaking exploration work and drilling wells near their land, that offered an immediate, tangible and personalised expression of all they hoped for. 'Company' itself was an abstract entity.[1] But in its physical and personalised manifestations it offered opportunities in the present to those who could discern and act on those opportunities. The camp at the airstrip, and the activities that took place there, were woven into everyday life. People watched the comings and goings, gossiped about things seen passing through or accumulating at the site, and speculated about what these activities might mean for future plans. 'What's the news from Company', people so often asked us and each other. A few gained access to the site itself, as cooks or cleaners, as labourers or security guards. These *kampani bois* (and very occasionally women) became brokers of information, and sometimes of goods and services. They passed on rumours, smuggled mobile phones in for charging and sometimes, as with the rice and noodles mentioned above, delivered largess from camp to community. They mediated relationships, not with Company as an entity, but rather with the individual people who represented Company; with individual people whose presence at the site was always short-term for turn-over of staff was the persistent state of affairs. The 'leviathan' that was Company, to use Golub's (2014) term, was mutable—changing form as the men who were its public face at Suabi came and went.

Sean was one of five white men who, through the first four months of 2014, arrived as camp manager before departing again for Australia, New Zealand or Port Moresby. Other men—Papua New Guinea (PNG) nationals—came as medics, mechanics or cooks. Helicopter pilots, or labour managers from the Juha camp passed through occasionally. More senior Company representatives visited rarely. The identities of these men, their personalities and practices were scrutinised closely, their names, origins, connections and behaviour discussed as local people clustered at the camp boundaries to watch what was happening. To establish at least the semblance of a personal relationship with one of these men—to be able to declare him 'my best friend'—was seen as key to accessing what the camp had to offer by way of either information or material goods.

These everyday interactions, and the access to benefits that they facilitated or constrained, are the focus of this chapter. They reveal much about how local people understood the nature of Company and their relationship

---

1   In this chapter, in accord with local practice and as an anonymising courtesy, we usually refer to the amalgam of companies utilising the Suabi camp as 'Company'.

with it. But these interactions were framed, too, by the understandings of Company representatives. And the understandings did not always coincide or mesh.

Taking and Asking

The acquisition of Company property occurred regularly throughout the early months of 2014. Building materials and tools were greatly favoured, with nails high on the list of preferred items. Men returning from work in the late afternoon or early morning would often carry a small, wrapped package of goods. They might, during the day, conceal what they had taken behind a tree or in vegetation near the river-bank and collect it when they finished work or, later, under cover of darkness. Foodstuffs were also popular, especially with men who, as trainee helicopter loadmasters, came from Juha to organise the shipment of supplies to the Juha campsite or were returning to Suabi because their work had terminated or they had been granted a break. Chicken noodles, biscuits, sugar, peanut butter and sweet soft drinks—taken from the apparently endless supply in the mess—were favoured items.

People were careful to minimise the likelihood that their actions would be seen by Company representatives and, for the most part, preferred that other community members were unaware of what was taken until after the fact. In the cases of tools and nails, secrecy was intended to reduce requests to share what they had obtained. But there was no sense of guilt, no sense that their acts of 'appropriation' might be judged as immoral. Indeed, with acquired foodstuffs, sharing was commonplace. When Sean gave weevil-infested rice to the missions, people came together for a community feast. The good-quality rice and noodles that were acquired at the same time were widely and publicly distributed, though then eaten at home or sold to others. Foodstuffs carried from Juha were openly shared, often to children who waited at the airstrip in the hope that their father would be coming home. We ourselves were the grateful recipients of a gift of two jars of peanut butter and, occasionally, were asked what we would like brought back.

There were, however, a few occasions when the community response was different. Late in 2013 someone wanting a length of hose seriously damaged the camp's water-supply system. This was a cause of much concern to Tassie, who was camp manager at the time. His annoyance was relayed to the community, which was always alert to the possibility that overt bad behaviour on their part might cause the resident company to depart. They found out who was responsible, raised PGK400 and, together with the missing length of pipe, delivered the money to Tassie as compensation. They did not name the thief. Tassie declined the payment; he did not know what he would do with it and felt that the bookwork that would be entailed in documenting the payment was not worth the effort. He suggested the money be given to the Mission, which could then use it to do something for the community.

On a later occasion, a large shipment of food supplies for Juha had been delivered to the Suabi airstrip. Men were sorting the delivery, readying it for transport by helicopter to the Juha campsite. A young man joined the workers. He was swinging a length of heavy saw-toothed chain attached to a handle. He picked up a carton of chicken noodles, shouldered it and walked away from the crowd to stand alone, watching from the edge of the airstrip. Men called him back. He did not respond. Someone ran to the camp to report what was happening to the camp manager. Company representatives appeared and also watched. But still the young man ignored everyone. A woman walked across the airstrip and talked to him. Eventually he gave her the carton, which she returned to the pile of supplies. Men had taken no action other than to call to the young man. They were concerned that if they approached him he might wield the chain, and the tension explode into violence. The woman who retrieved the carton was his mother.

Visiting companies regularly extend goodwill with gifts to the local community. In the late-2000s, Oil Search and ExxonMobil provided small backpacks and T-shirts to school children. The former carried the words '*wokim wantaim*' (working together) and the company logo; the latter, the words 'PNG LNG' and '*mi stap seif*' (I am safe). In 2013, ExxonMobil gave many wakawakas—small, solar-powered LED-light lamps that can also be used to charge mobile phones—for distribution to school children to aid studying at night.[2] Some reached the children but many were acquired by adults for personal use and, in some cases, including by schoolteachers, for sale. ExxonMobil also gave many copies of a specially commissioned six-book series of readers that followed a boy named Toea as he travelled through different parts of PNG and learned

---

2   Ben Coxworth (2013) provides a description and review of the wakawaka power solar lamp and device charger.

about the ways of life that he encountered. In 2014, in response to ongoing written requests on behalf of the missions, community health centre and elementary schools, several of the expatriate camp managers made a variety of small contributions (for example, fuel for the airstrip mower, rice and noodles to feed people who had worked to maintain the airstrip, some medicines, stationery, building materials), though they themselves were awkwardly placed with respect to the ambiguity entailed by the official need to obtain advance permission from their employers. There was the added complication, too, that while the camp was set up and operated by one company, most of the property held there belonged to another. Guy refrained from handing out Company property. He, however, had personally brought to Suabi hundreds of school readers from the 1960s and 1970s that he obtained in his home town after advertising in a local newspaper. With our assistance, he organised a morning assembly during which he shook hands with all elementary school pupils—none of whom could read but many wearing *mi stap seif* T-shirts—and gave each of them two booklets (Fig. 7.1).

Figure 7.1: Gift giving at the Company camp.
Source: Photograph by Peter D. Dwyer, 2014.

With one exception, it was rare that local people would ask Company representatives directly for goods or other forms of assistance. The exception was references that were sought when a person finished a period of employment. Many people carried such references. They came in three types. Some were from Papua New Guinean companies that provided labour, security, catering or other services to support exploration activities. These tended to take a standard form and different companies plagiarised each other's wording. Variants on the following were usual:

> During this duration, Mr. […] was found being matured, reliable and hard working person. He posses leadership qualities and executed given tasks under minimum supervision.

Other references were provided by officers of international companies. A few of these carefully outlined the kinds of work that the person had done and were informative with regard to that person's capabilities and competencies. Most, however, said no more than that the person in question had indeed worked for the company between stated dates. The third category comprised fraudulent references on which the name of the intended recipient had been deleted and the name of a different person inserted. Because the computer skills of the forger were seldom of a high order it was nearly always obvious that the reference had been manipulated. We ourselves were often asked to alter a reference in this way. Our refusal to do so, and our discussion of fraud and its potential consequences, were understood to be our own peculiar idiosyncrasies. People knew that in the modern world references were useful when seeking employment and presumed that, therefore, many people wrote their own. In the one case where we ourselves wrote a very favourable reference for a young man who had worked for us, we learned a year later that his Port Moresby-based 'brother', whom we had not met, had found it 'very helpful' in his own endeavours to find employment.

When seeking assistance from companies people preferred, and for the most part adopted, indirect contact. This was a practice of long-standing that applied more often when approaching people at the fringe of, or outside, local social networks than people within those networks. With insiders there was little ambiguity. A person would not make a request unless confident that there was little likelihood of refusal. With outsiders, however, there was the risk that a request would be denied; it was difficult to know in advance. And a denial would be shameful to the person making the request. In our earliest visits to Gwaimasi we provided a great deal of low-level medical assistance: treating cuts, ulcers, fevers, infections, skin fungus, toothache and so forth. For a long time, people wanting medical attention either came with someone who made the request on their behalf or said, for example, '*ma odio do igidehai*' ('I have a very bad headache') or, less attractively from our point of view, positioned their often suppurating wound directly in front of us and waited for us to offer appropriate treatment. We had offered; they had not asked. At Suabi, too, during our more recent visits, people who we knew less well adopted similar strategies. A man would arrive carrying paper, holding it in such a way that we saw it, and wait until we offered to type the handwritten words. Some delegated another person to ask on their behalf.

Letters provided a new way in which people could distance themselves from the request they were making. The 'request' was in the written words; it had not been spoken by the person who hoped that the request would be fulfilled. And, further, to reinforce the distancing that was sought, it was seldom the case that the person making the request would be the person who delivered the letter. Company representatives received letters requesting assistance written on behalf of the mission, community health centre and schools. Other letters came from individuals seeking employment, or free travel on Company-chartered flights that had delivered supplies to the exploration camps. Kigi, for example, had been employed as an assistant foreman on the Juha seismic project. He worked for ten weeks without a break. In the closing weeks of that period he was based at a bush camp at high altitudes in Hela Province, close to Levani valley. Armed Huli-speakers from the valley visited the campsite, challenging Western Province workers and threatening harm if they did not leave. Understandably scared, Kigi left the field site but, under then prevailing practice, his departure meant that his employment was terminated. A few months later, he learned that a new seismic line was being cut and that this was distant from areas populated by Huli-speakers. He wanted to return to work but needed to explain why he had previously abandoned his position. In his letter, addressed to three named camp managers at 'Juha Mining Camp', Kigi wrote:

> When I was working on Lines (4) to Line (3) I worked for two months. While I was working Lebanny [Levani] people came to the campsites bringing bushknives, axes and guns. They made us very frightened to work for money in their land. With permission I came back to my home at Suabi Station on December 29th, 2013.

In January 2014, young men wanting to return to schools in the Highlands wrote letters to 'Oil Min Field Services' and the 'Gama Camp Manager' requesting free flights from Suabi to the Highlands. The first of these letters requested lifts for 16 named males 'because this is the good relationship between your Company, youths and the communities'. There was no response to this letter. The second, to 'Gama Camp Manager', requested lifts for 12 named males, asserting that they were residents of Suabi, Siabi or Gesesu 'where the project is operating now' and stated that:

> We the total of (12) students attending High Schools, Secondary Schools and Technological College at Highlands Regions are seeking for transport from Suabi to Mendi by South West Air during next week from your charter backloading.

Eleven males were eventually provided with a free flight to Mendi though, in fact, none of those listed was a resident of either Siabi or Gesesu, some had not been named in the letter and some who flew did so for reasons other than furthering their formal education. This proved to be the last occasion on which a locally based company provided free transport. Talisman had decided that, at 650 m, the Suabi airstrip was shorter than required for safe landing by Twin Otters and, thereafter, began ferrying supplies and people by helicopter from the longer airstrip at Nomad.

In only one case did we see and, with some improvements to expression, type a letter of demand. This was addressed to 'Juha 4 Camp Oil Min Field Service Ltd', was to be signed by six 'owners of the ground on which Juha camp has been built'—men from two Febi clans—and asked the company to sign a new agreement for 2014. The letter stated that 'we would like your Company to appreciate the difficulties that we, the owners of the ground, are facing' and asked that company workers receive free transport to Mount Hagen and Kiunga, sick people be airlifted to hospital and school students receive free lifts to towns at which they would attend schools. The letter included a list of revised rental charges for camp house areas, lay down area, rig board, helipad, water pump, generator and fuel tank sites and 'all other buildings' and stated that 'at the time the Company moves away from the Juha camp, items that are to be abandoned must not be burned or destroyed but must be given to the owners of the ground'.

The total amount of rent sought was PGK5,000 per month.[3] On the day the letter was carried to Juha by someone returning there for work, a man purportedly representing the two Febi clans spoke aggressively to the helicopter crew. None of the signatories to the letter spoke. There would be trouble, the speaker said, if their demands were not met. News of this altercation reached Juha. The next day Mick, the current Juha labour manager, visited Suabi. He asked a group of Kubo people about the previous day's 'problem'. They were reticent, unwilling to admit any knowledge of what had happened. Mick pushed for details. Who was it, he asked, had harangued the helicopter crew? Reluctantly, people gave him the name, making it clear that the person concerned was a Febi man from Gesesu; he was not one of them and they were not responsible for his behaviour or the demands that had been made. Mick knew the man and laughed the incident off. '*Em i yunpela man, em i no lidaman*' ('He's a young man, he's not a leader'), he said sarcastically. The protest and letter of demand had no effect.

When Company Came

The flow of goods from the camp to people at Suabi—whether simply taken by local people or demanded by them, gifted by Company representatives or provided in response to requests—was substantial. To understand what was going on in these transactions, however, it is necessary to look at how the relationships that underpinned them were negotiated.

Companies operating in the territories of Kubo and Febi people make numerous contributions to the communities where they establish campsites. They do so in the name of the company and as an expression of their social responsibility. Oil Search, for example, reports that:

> [Our] approach to sustainable development is collaborative, relying on the formation of close relationships with key stakeholders such as host communities, governments and our development partners.

---

3   At Suabi, in 2013–14, Talisman paid PGK50 a day as rent—approximately PGK1,500 a month—for the entire campsite and the facilities located there. At the Fasu community of Haivaro, in Gulf Province, rental arrangements associated with a 2008 Talisman and Sasol Petroleum seismic operation were assessed on the basis of separate payments for campsite, water pump site, water source, airstrip landing fees and helicopter pads and amounted to a total of PGK2,900 per month (Fitzpatrick 2010: 63–4). Haivaro people have had decades-long experience with the timber industry and were likely to be attuned to possible returns and experienced at negotiating those returns. At Suabi, people did not negotiate but took what was offered.

> Built on mutual trust, these relationships help to ensure the Company's development programmes are effectively targeted, efficiently managed and continuously improved.[4]

This company publishes booklets that illustrate ways in which it works 'with local people to make a difference to their everyday lives and build a sustainable future for their children and their communities' (Botten, in Stone 2012: 1). In PNG the human dimension is clearly evident in the ways in which oil and gas companies communicate with a broader public.

The men on the ground—camp managers, labour managers, etc.—also seek to establish good relations with local people, though their largesse is constrained by the facts that they are beholden to distant employers, are charged with minimising costs and, to varying extents, confronted by incidents of theft, damage to property and, sometimes—though not at Suabi—outright aggression.

There is, however, another side to the interactions between exploration companies and the Suabi community, concerning what is not done rather than what is done. Our interest here is with the behaviour of representatives of resource-extracting companies 'on the ground'. How do they engage with local people? How do they fulfil obligations to their employers and to what extent do they meet expectations with regard to social responsibility at the time they initiate a project on the land of other people? This other side of interactions may be illustrated by events surrounding establishment of an exploration camp at Suabi late in 2012.[5] Those events reveal arrogance on the part of Company and complicity on the part of the local community that, taken together, since these behaviours are not entirely unexpected, are more deeply indicative of strategies whereby each of the participants navigates the vastly different social and material worlds they encounter but do not understand.

---

4   Viewed 28 February 2017 at: www.oilsearch.com/__data/assets/pdf_file/0009/7749/2015-Oil-Search-Approach.pdf.
5   It is not always clear to outside observers which companies are associated with a particular project. The primary operator may subcontract others to undertake particular components of a project and the latter may, in turn, subcontract cooks, medics, labour force, transport, etc. In 2013, for example, a minimum of eight companies were represented in the work force at the Suabi camp.

Arrival

'Company is coming', people called. As usual they had heard and seen the helicopter before we did. The helicopter had been chartered by an oil and gas exploration company. Three men, the senior man 'white' and the others Papua New Guinean, had come to inform local people that they planned a laydown—an exploration base camp—at the village. They wanted to rent the Mission House—the US missionary and his family had departed eight years earlier—for the duration. They said that the camp would be small and they would stay for only six weeks. The people at the village were excited. They would be paid to build and service the camp, they would be paid to cut 'lines' through the forest and, if valuable resources were found, they would receive royalties.

The crowd that met the helicopter was dominated by women and children, because most men, youths and village dogs were in the forest hunting for a forthcoming feast. Alex, the community health worker and himself a local man, had obligations at the village and had not gone hunting. He immediately took charge of negotiations with the three arrivals—they had quickly checked that we were not missionaries—and led them to the Mission House. He did not tell them that the house was currently occupied by the pastor, Martin, and his family, though it would have been obvious that people were living there. Martin himself was away hunting.

The visit was brief. The Company representatives inspected the house, held a meeting inside with local men—the house was high set and the women sat underneath in the shade—marked out areas where they planned that two large marquees would be built to house workers, identified positions for a laundry, ablution block, toilet and generator site, checked water tanks and the solar system, and said that they wanted 400 poles cut and ready to commence building on 5 January. There was no discussion of payments they would make to rent the house or for poles. There were no details about where the exploration lines were to be cut or the number of people to be employed for this work. They said they would return on 20 December—one of three planned feast days—to continue discussion about their needs.[6]

---

6   We did not join any of the meetings that took place inside the Mission House. They were of great interest to us but were not our 'business'. We were not alone in making this decision. Two men—one from Oksapmin, the other from Morehead—who had, in the past, been community health workers based at Suabi and had married local women, made the same decision for the same declared reason.

Men, youths and dogs arrived from the forest through the next few days. With the exception of the community health worker, all men holding 'official' positions in the community had been absent when the helicopter arrived, either hunting or at Kiunga. It was some or all of these men that the Company representatives should have consulted. Pastor Martin told us that his wife had been 'disappointed' by the behaviour of the men who had inspected her house without consulting her. She had sat with other women under the house during their visit. Martin himself said that Company men had failed to show 'respect'. He confirmed with others that no rental contract had been signed.

Through the next week, some men and youths commenced clearing the ground where marquees were to be built and others travelled out from Suabi to cut poles, float them down the river, and stack them in readiness. Through the same period, however, people from other Kubo villages and from neighbouring language groups started to arrive. They were invitees to the feast, became caught up in the excitement of what might be happening and, like most local residents, hoped that they too would soon be contracted for the planned Company work.

Company men returned on 20 December, the second day of the feast when 65 domestic pigs were exchanged, killed, butchered and, through the night, cooked in heated-stone ovens. On this day three men arrived. They were all Papua New Guineans with one, Taylor, a community affairs officer. Pastor Martin, with supporters, met them on the stairs to the Mission House. He was blocking their path. He spoke strongly to the Company men and the large group of local people. The area of land with the Mission House and associated buildings, he said, had been given to the ECPNG by local people, and the 'things' on this small area belonged to the mission. He asked those who were listening whether they agreed. There were no dissenters. In this public forum, Martin made it clear that, with respect to the Mission House, Company should negotiate with the mission—with the pastor and church leaders. He did not say—though all local people understood his intent—that the community health worker had no authority with respect to allowing people to use the Mission House.

When Martin had finished speaking, Taylor addressed the assembled group. He spoke in a mix of English and Tok Pisin with Alex translating into Kubo. The planned project was small, he said. Lines would be cut through the forest and, when this had been done, geologists would arrive, walk the lines and survey the ground. There would be no seismic testing. Company had come to Suabi because this village provided the most convenient access to the area of exploration. They would employ landowners to cut the lines but, as yet, did not know who these people were. Taylor stressed that, at this juncture, nothing had been found, and it was possible that nothing would be found. People should not expect that they would necessarily receive 'big money'. That depended on what, if anything, the survey found. He handed over one copy of a map, printed on an A4-sized sheet of paper, showing a small portion from a topographic map on which the proposed survey lines had been drawn. He did not tell people where the area depicted was located.

There were questions and comments from the audience. One man suggested that Company had come to Suabi to establish their base camp because Huli people, to the northeast, were too demanding. He was correct, but his comment did not elicit response. One of us asked whether a social mapping study would be carried out in advance to identify 'landowners'. Taylor replied that this knowledge emerges slowly and is subject to revision as the project continues. Several people stated that there were no identifiable rivers or mountains on the map that had been provided and, therefore, they did not know where the licence area was or where the lines would be cut. No explanation was provided.

The Company men and some local males now moved into the house to discuss financial and other administrative matters. They offered PGK500 to rent the Mission House for the period that they would be based at Suabi. But Martin had come prepared. He had drawn up a price list that we had typed and printed on his behalf. For businesses, such as oil and gas companies, rent was set at PGK120 per day with a reduction to PGK720 per week for long stays. His document stated that 'access to radio room and store room where battery is held by Mission staff and by MAF [Mission Aviation Fellowship] staff must be maintained at all times' and that there was to be 'no access to radio room by guests at any time'. (The radio provided contact with mission-operated medical and airways services, and the room in which it was kept was attached to the house with entrances from both inside and outside.) No agreement was reached. Martin's demands, he was told, would need to be discussed with the 'bosses' at Company's home base.

We photographed the small map that Taylor had shown, and matched it to one of our topographic maps. The proposed exploration area was 45 km north of Suabi, approximately 25 km northwest of the existing Juha gas fields and in an extension of the same geological formation (see Fig. 3.2). The combined length of the base line and three cross-lines was 45.95 km. Most was in the recently named new province of Hela; only 14.4 per cent of the total was in Western Province. The implications were clear. No Kubo person would qualify as a landowner, and of Febi it would be, at best, only a few with affinal connections to northern language groups who might justifiably assert rights to that land. Company would have known this from earlier social mapping studies in the region, studies that they themselves had funded (Ernst 2008; Goldman and Ernst 2008), but had withheld information. They were disinclined to operate from a base in Huli territory and intent on ensuring the cooperation of Suabi people (compare Warrillow 2007: 103–4). To that end, holding out hopes of employment and future benefits seemed wise.

Company came again, four days earlier than expected, on 1 January 2013. This time, three Papua New Guineans and one 'white' man arrived. Mick had 40 years experience in PNG, and at once established a warm rapport with local people. We told him of people's concerns about the lack of respect shown to the pastor's family on the first visit by Company. At a later meeting inside the Mission House—again, with only males in attendance—Mick apologised for earlier behaviour and said he would raise the matter at headquarters. Martin discussed rent and reported that Mick had responded favourably to the prices he proposed. But no contract was signed; he and Mick shook hands and talked of 'trust'. Mick left soon after the meeting. But the three Papua New Guineans stayed behind. The campsite was now in place and these men were tasked with converting it into an operating base. The senior man was Dave. He had come on all three visits and was second-in-charge to Ronald, the 'white' man who had come on the first visit and was to be camp manager.

Through the next two weeks seven local men were employed to build and fit out the two marquees, and another as 'security officer'. They were supplied with overalls, boots and dark glasses and, each morning, before work commenced, attended a meeting at which 'safety' issues were discussed. One of the marquees provided cover for eight large one-man tents, each with a bed, small cupboard, lamp and fan. At the entrance to each there was a 'welcome' mat. It was here that more senior Company people would sleep when the three-bedroom Mission House was fully occupied. A storeroom was built beneath the Mission House, and almost every day helicopters arrived to offload food, a refrigerator, gas cylinders, a pump, building materials, safety and other equipment, and drums of fuel. A generator was installed and, thereafter, through the night, the campsite was flooded by security lighting.

Before long, tensions began to emerge. More poles were needed to provide frames for the laundry, ablution block and toilet, but poles already delivered had not been paid for. Timber that had been stored beneath the Mission House—the property of the Church—was used without asking permission. The trunks of two coconut palms were wanted to frame the toilet pit. Pastor Martin offered two palms that were not producing. Company men felled one of these and, without asking, another that was producing well.

By 14 January the entire area including the Mission House, marquees and ancillary buildings had been enclosed by tape that was striped in red and repeatedly featured the word 'danger'. Access to the radio room was not closed off but the tape sent the unambiguous message that people were not welcome and, further, closed off the area beneath the house where people sheltered from rain and sun while waiting for mission planes or medical information. Local people were irritated by the presence of the tape, insulted by use of the word 'danger'—was it them, they wondered, who were thought to be a danger. Quite regularly, people (often women or children) leaned against the tape until it snapped and then, because the barrier it represented was no longer present, moved into the shade.

Rumours continued to circulate concerning the number of people who would be employed to cut lines, the number of these who would be from Suabi, and the day when the list of employees would be drawn up. On a few occasions arguments broke out when men, anxious about employment opportunities, challenged others to justify their status as 'landowners'. But an increasing number of men were beginning to doubt that the tally of employees would be high or that many of those who were employed would be Kubo-speakers.

On 17 January, Ronald, Taylor and a contracted cook returned and would now stay for the duration of the project. Ronald and Taylor stopped briefly before entering the Mission House. They were positioned inside the 'danger' tape and, from here, talked to the assembled people. For the first time they acknowledged that the lines to be cut were partly in Hela

Province and partly in Western Province. The work would need to be divided: to be split *namel* (in the middle) between people from the two provinces. They did not acknowledge that less than 15 per cent of those lines was in Western Province. Taylor said that he would meet with people later that day to make plans. He told them to draw a map showing villages near the exploration area so that landowners who should be consulted could be identified. He stressed, once more, that the job was a 'small one', that it was a geo-survey, not a seismic survey. Several men in the audience muttered that this indeed meant little work would be involved though, at this juncture, they were left with the impression that at least half the workers would be from Western Province.

By 2 pm people had assembled outside the Mission House. Febi people predominated; Kubo people were beginning to accept that they had no claims to land in the exploration area and would not be employed to cut lines. People were showing signs of frustration because Taylor had not appeared. Horace—a candidate for forthcoming district-level elections—spoke strongly: 'It is nearly three o'clock and no one has come to talk'. Another man interjected that people had been waiting since 1 January and there was still no list of people who would cut lines. Horace reminded everyone that Company men had taken over the Mission House, suggested that they would be drinking and associating with women, and said that such behaviour was inappropriate in 'God's House'. He asserted that he himself was a 'border man' and should definitely be consulted.

A local man went into the house to express people's frustration at the delay. Taylor appeared and sat on the grass. He had brought a rough map—a pencil sketch. He spoke of the need to identify villages near the exploration area on both sides of the border, and then to find landowners and distribute work. He could not provide a better map, he said, because the printer had not yet been set up; everything would become clear when a proper map could be provided. Some Febi men had also prepared a sketch map on which they had named some villages on both sides of the border. In a few days' time, Taylor said, some leaders from Western Province, and others from Hela Province, would be flown across the exploration area to clarify issues of landownership. He stressed again that this was a small job, designed to facilitate access by the geologists. He said that only three crews, each of ten men, would be needed—in fact there were to be five crews—but added that some Suabi people would be employed to do the laundry. He then talked of the 'difficulties' created by Highlanders and said that it was for this reason that Company had come to the 'cooperative' lowland people.

The helicopter trip to the north was scheduled for 19 January. By this date Horace had been provided with a laminated copy of the small map seen weeks earlier. He showed it to others waiting at the Mission House. '*Dobai kauimi*' (not big work), he explained. And continued by asking rhetorically why, given that the work was in the Highlands, Company had chosen to make their camp here at Suabi. Three men from Suabi—two Febi and the village councillor who had returned from Kiunga in time for the December feast—were taken on the flight to the north. Others were picked up at Tobi, the northernmost of the larger Febi villages, and from Geroro, a Duna-speaking community approximately 12 km east-northeast of the exploration site on the Hela side of the border. The Tobi participants stated that they had some connections to land on the Hela side but did not assert strong claims. The Hela participants asserted that they had full rights to all land in the exploration area on both sides of the border. The trip did not resolve questions of landownership. Both Company people and local people often spoke as though provincial borders matched language group borders, but on the ground, or when actually viewing the ground, local people brought understandings of past and recent engagement and of marital connections to bear on their interpretations. It was rumoured by some that when Hela people viewed the land, seeing that it was exceptionally rugged and dotted with limestone sink holes, they were afraid and said they did not want to work there; if the lowlanders wanted the work they could have it. This was wishful thinking and the rumour did not gain currency.

By this time a 'medic' had arrived and a clinic was being built under the Mission House. The medic was to conduct clearances for all the men who would cut lines, and deal with emergencies if any occurred, but would not himself be based at the field site. Company had contracted Alex to fill the field position. Ronald rationalised the fact that this would leave Suabi without a community health worker by telling us that Alex was eligible for two weeks annual leave. He did not comment on the fact that cutting lines would take much more than two weeks.

More poles were cut and delivered for building purposes, and people were employed to carry stones from the river to the campsite to lay paths between the Mission House and the marquees, laundry, toilet, etc. Company employees could not be expected to walk through mud—rainfall is high at Suabi—when moving between locations.

From 21 January the contracted helicopter was usually based at Suabi because it was expected that, very soon, men would be shifted to and from the field sites. There were delays, however, because wireless communication between the field site and Suabi was problematic; Suabi could receive messages from the field site but the latter was not receiving Suabi. The field project could not be initiated until this problem with the repeater station was resolved.

On 23 January another white man joined the team—Kane, a petroleum geologist, who had been contracted to, as he put it, 'hold the fort' during times when Ronald was overseeing work at the exploration area. Kane was to be responsible for ensuring that communication was maintained between the remote mountain camp and the Suabi base camp. He had been flown from Wales to take up this appointment. He told us that the 'job' should be finished by the end of February, in six weeks, which was the original estimate of the duration for the planned project. He told us also that 30 men would be arriving soon. They are 'foremen', he said, employed by Company and experienced at cutting lines. In addition, there would be 30 'labourers'. We understood now that each of the so-called foremen was to work with a labourer and that only the latter would be selected from among recognised landowners. Once more, the number of local people who might be employed had been, publicly, whittled down.

On 25 January Company was obliged to change arrangements with regard to water supply. Water was needed both for the Suabi base camp and for delivery, by helicopter, to the field site. The Mission House tanks could not meet the demand. Dave, who had been left in charge of establishing the camp, had planned to pump water from the Baiya River and a pump had been delivered for this purpose. It rained heavily before the pump could be installed. The river flooded—it became a raging torrent—and it became clear that the initial plan was seriously flawed. Dave then decided to take water from a tank 500 m from the Mission House. Without asking, he got the workers to lay a narrow pipe from that tank to base camp. Local people were deeply concerned. This tank was the primary water supply for more than 250 people who lived in two of the clusters of houses that make up Suabi. They stressed that it had recently been a 'dry time'—there had been little rain—and their own needs for water were not being considered. They prevailed upon Dougal, the Suabi Council, to talk to Ronald, who had returned to take charge of the now-established camp. It was then decided to obtain water from a large and fast flowing stream on the other side of the Baiya River. Drums of water—one at a time—were to be transported by helicopter.

Through the next few days Company provided free helicopter lifts for nine youths and one young woman who were departing for the remainder of the year to continue secondary school studies, or in one case tertiary-level technical study, at Kiunga, Tari, Mount Hagen, Goroka and Lae. The students would be dropped off at places where Company had major facilities and would then make their way by foot and public transport to their target locations.

On 25 and 26 January the 30 'foremen' arrived, though the ongoing failure to establish radio communication between Suabi and the field camp meant that line cutting could not yet be commenced. The foremen were accommodated in one of the marquees, received some training in loading and unloading the helicopter, and waited. On 27 January we learned that Company now wanted full and exclusive access to the Mission House. They did not want local people waiting around the Mission House. They asked Pastor Martin and the radio operator to move the radio from the room where it had been installed 17 years earlier to either a very small storage room under the house or, preferably, to another location away from the house. Their request was denied.

On 28 January we left Suabi for five months. We spent much of that morning near the Mission House, waiting for the plane. To this time no money has been paid for rent or for the poles that had been cut for building the camp. Some people from outside Suabi had contributed to the latter work and remained in the village waiting for payment, being fed and accommodated by local people. Two young women had now been employed to wash laundry, and two additional local men were being trained as helicopter loadmasters. The communication problem between Suabi and the field camp persisted, and none of the foremen or labourers had yet been moved to the field camp. Five weeks had elapsed and the project was still not off the ground. We asked Ronald where the 'land-owning' labourers would be selected from, but the most he would say was that they would be a mix of Hela Province and Western Province men.

He volunteered, however, that it was difficult working in the Highlands where you were likely to 'get a gun pushed in your face' and, by way of self-congratulation, told us that Company had facilitated free trips in the direction of school for quite a few Suabi students.

Aftermath

In early July 2013 we returned to Suabi for one week.[7] Through February and the first three weeks of March, the survey lines had been cut and the geo-survey was completed. The hired labour force comprised 17 men from the Duna community of Geroro and 13 from the Febi community of Tobi. The three geologists who undertook the ground-based geo-survey were not Tok Pisin speakers and were assigned assistants who were relatively fluent in English. One of the assistants came from Geroro, the other two—including Martin—were Kubo-speakers from Suabi. On the final day of the geo-survey the three parties—each comprising a geologist and his assistant—were lifted by helicopter to Geroro to pick up rock samples they had accumulated through the past week before returning to Suabi. They found the samples being guarded by Geroro men who challenged the geologists, asserting that no environmental impact report had been conducted and, thus, the geologists had no right to remove resources from Duna land. The encounter grew increasingly heated. The Geroro employee attempted to leave the helicopter but was restrained by the two Suabi men who considered that their own safety was enhanced if at least one Duna man remained with them. On Martin's very strongly worded advice the helicopter pilot departed for Suabi without the rock samples. At Suabi, Ronald, the camp manager, ordered Martin to return the next day to collect the samples. Martin refused, saying that 'life' was more important than 'stones'. His employment, which he had found very interesting, ceased. The rock samples were eventually collected, though we did not learn what negotiation was entailed at Geroro.

The Suabi camp was now closed. Company paid Martin PGK100 per week for the 11 weeks they had occupied the Mission House. This was a fraction of the amount that he had thought had been negotiated and that had been stated unambiguously on the price list provided, on separate occasions, to both Taylor and Mick.

---

7   Our account of Company activities from February through June 2013 is based solely on reports by Suabi residents.

Company left Suabi and then returned a few days later on 28 March. It had been decided to undertake a long-term seismic survey in the Juha area. They wanted to rent the Mission House again but Martin refused, saying that he was using the house to run a year-long Bible school for a number of Kubo, Febi, Samo and Bedamuni men and women. Local people understood that he had been disappointed by the amount received as rental payment, and several told us that it was for this reason that he refused Company's request. Through April, a large campsite was established near the airstrip on land that had not been assigned to the mission. Through May and most of June, a contingent of 'foremen' was brought to Suabi and many men from local language groups were employed. In late June a serious helicopter accident in the mountains was attributed to bad weather, and the seismic survey put on hold until the weather improved.

In early July, the campsite was in the care of two Company men (a camp manager and a mechanic), two local men acted as night-time security guards, a few others as labourers and Alex had been contracted to oversee any medical needs of people remaining at the camp. Six Febi men were based in the mountains, two to provide security and maintenance at a repeater station and four to provide security and maintenance at a field campsite. Though the seismic survey at Juha had been suspended, helicopters came to Suabi on an almost daily basis. They were servicing social mapping and environmental impact teams along the route of a different proposed gas pipeline to the west of Suabi. This route had been chosen to minimise engagement with Huli in the Highlands.

More surprising to us than all this was the state in which we found the Mission House. That house had been built in 1986. It was high set. On the advice of Tom Hoey, the long-term Bedamuni missionary, the stumps were of an extremely hard and naturally insect-resistant wood, and were oil-treated and capped with metal to further minimise access by termites. In early 2013, the house was still in excellent condition. At that time, Company built a storeroom and a clinic in the area under the house, with more than 40 timber uprights framing the two rooms. But none of those uprights was protected from termites, and none was removed when Company departed. Had they remained in place they would have provided easy access for a termite invasion. And, further, when Company used the Mission House early in 2013 they disconnected the solar pump that fed water to the header tank and, instead, used their own generator. When they departed they took the generator, failed to reconnect the solar pump and misplaced required fittings.

Through those first months of 2013, then, it seemed that Company representatives paid little attention to the needs or concerns of local people. At the same time, local people went out of their way to minimise the likelihood that they might be perceived as difficult, as uncooperative. Though there were certainly grumbles behind the scenes, they did little that could be judged as 'resistance'. Company representatives were 'arrogant'. As Martin had put it, they did not 'respect' those with whom they engaged; they did not acknowledge local people as persons. Local people, however, were complicit, always attending to signs of what Company wanted and adjusting their presentation to suit.

Arrogance and Complicity

In a far-reaching analysis of controversies surrounding the Jabiluka uranium mine in Australia, and of the position of indigenous people vis-à-vis those controversies, Subhabrata Banerjee (2000: 4) argued that 'current organization theories on managing stakeholders are complicit with colonialist attitudes and values'. The problem, he argued, is that 'social appropriateness'—a catch-cry of those theories—is often 'subsumed under notions of "progress" and "development"' (ibid.: 24) with the outcome that the interests of indigenous stakeholders are, ultimately, subordinated to those of the mining company and the state. Those notions are substantiated through capitalist and colonialist discourses that prioritise science, the market and assumptions of superiority–inferiority, and assert a morality under which their concomitants are assumed to be beneficial to all stakeholders. Such discourses are, ultimately, self-righteous and arrogant.

Our assertion of 'arrogance' on the part of Company representatives at Suabi is based not on what they said but, rather, on what they did not say, or what they delayed saying, and on what they did. It was five weeks after Company arrived at Suabi before it was acknowledged that the exploration area fell well outside Kubo territory and, indeed, that most of the area fell outside Western Province. The first map they showed was uninformative, and the next a mere sketch. Only after six weeks, when the 30 foremen arrived, did it become clear that very few local people would be employed to cut lines. Throughout those early weeks, Company representatives withheld information that, from their decades-long experience in PNG, including in the Juha area, they would have known: for example, that Kubo territory was separated by one language group from the exploration area. Not until the base camp was thoroughly established were they more forthcoming about their plans and intentions. In all these ways, Company failed to adhere to well-established guidelines of 'free, prior and informed consent' that are intended to promote responsible interactions with indigenous people (Whiteman and Mamen 2002; Salim 2004; Macintyre 2007; Gilberthorpe and Banks 2012).

But the arrogance of Company representatives was most evident in practice: for example, failing to activate or fulfil contractual obligations with respect to renting the Mission House; taking water from a tank that provided the needs of more than 250 people; contracting the local community health worker to relocate to the field campsite without due consideration of implications for the health of village residents; and attempting to have the radio moved from the room it had occupied for 17 years. Moreover, in early meetings with local people, Company representatives said that they had come to Suabi because it was most 'convenient' to the proposed exploration area. The impression was that Suabi was closer to that area than any other accessible location. This reinforced the (mis)understandings that the survey lines would be cut on Kubo land, that Kubo people would be employed to cut lines and that, as landowners, Kubo people would be eligible for royalties if gas was found. Not until mid-January, with the camp fully in place, was it acknowledged that the choice of Suabi had been made to minimise engagement with 'demanding' Highlanders; lowlanders, in contrast, were 'cooperative' and thus more desirable hosts.

When scholars invoke 'complicity' it is often in a frame of disparaging the efforts or understandings of Western thinkers or actors who analyse, or engage with, the lifeways of people thought to be, in some way, at a disadvantage either relative to the 'West' or relative to other members of their own society. Anthropologists or feminists, for example, though asserting defence of the disadvantaged, may be charged with being complicit with prevailing Western discourses that are seen as the ultimate source of that disadvantage (McPhail 1991). There is less inclination to report and analyse the ways in which purportedly disadvantaged people may be themselves compliant with that which befalls them and, in consequence, complicit—perhaps innocently so—in the ideological persuasions of their purported oppressors. Rather, there is a bias for demonstrating ways in which those people 'resist' the impositions of the world beyond themselves (Ortner 1995; Brown 1996; Seymour 2006). There are important exceptions, particularly under the rubric of 'symbolic violence' (Bourdieu 2001). In some analyses, for example, women are reported to be, at least in part, complicit in their own purported oppression by men (Strathern 1988; Moi 1991). Complicity and resistance, however, may be closely connected and, in different circumstances, either may give way to the other.

At Suabi, desire for the actual and imagined benefits of development, combined with understandings of the authority and expectations of both the state and state-sanctioned industries, underlay the compliance of residents with and, ultimately their complicity in, much of what happened when Company arrived to establish a base camp on their land. In usually performing as asked or expected, they were aligned with the motives and ideological persuasions of Company irrespective of their understanding of, or agreement with, those motives and persuasions.[8] This is evident throughout our account of the laydown. A striking, and oft-repeated, instance was the spatial and participatory segregation of women and men that occurred at meetings with Company representatives. Women never joined a meeting that was held inside the Mission House but, rather, assembled beneath the house waiting to learn of outcomes. And at meetings held outside they stood, or sat, apart from the men and did not contribute to discussion. This segregation was in marked contrast to usual practice at community gatherings; women and men mix and converse freely at events such as local court cases, working bees and informal sport. The one exception was at church services where, when these are held in the church itself, women and men sit on different sides of the building. When the church building collapsed and, for some months, services were held in people's houses, there was no spatial separation. The gender equality

---

8   While acknowledging that relationships between Kubo people and Company could be analysed in terms of symbolic violence—'violence wielded with tacit complicity between its victims and its agents, insofar as both remain unconscious of submitting to or wielding it' (Bourdieu 2001: 246)— we are disinclined to unilaterally declare Kubo 'victims', preferring instead to see much of what they did as strategic.

that usually prevailed among these people was voluntarily suppressed in circumstances where it was presumed that intruding, and powerful, outsiders had quite different expectations of roles.[9]

Other expressions of complicity included the undertaking of work with no prior arrangements concerning remuneration and the failure to indicate disagreement with Company activities to Company representatives, with expressions of concern being confined to discussion among members of the local community. Again, it was striking that local people employed by Company acceded to all demands even when they knew that the required tasks would inconvenience or disturb other community members. The employees informed community members that this work was underway, but never themselves challenged the instructions they received. Nor did community members rebuke the men who carried out these tasks; they accepted that the employees were beholden to Company and understood that a person who complained could be easily replaced.

People at Suabi were often disgruntled by the activities of Company and by Company's failure to provide relevant information. But this seldom spilled over as resistance. People talked among themselves or made rhetorical statements that implied dissatisfaction, but their acts of resistance were for the most part minor. They were motivated to act, after the fact, when it became apparent that Company intended to draw substantial quantities of water from a community tank. More striking, perhaps, was the frequency with which people 'accidentally' broke the tape that was intended to mark out an area from which they were excluded. There was no prior discussion about doing this.[10]

On one occasion, Pastor Martin, with supporters, delayed access to the Mission House by Company representatives until he had made a speech to assembled people. On other occasions he refused to comply with a request to relocate the radio, refused a demand that he return to Geroro to collect rock specimens and, when Company returned a second time, refused their request to again rent the Mission House. On none of these occasions, however, did he indicate anger. He conformed to the local desire that Company be made to feel welcome and should neither be

---

9   Martha Macintyre (2011) discusses ways in which females may be disadvantaged within the context of mining in Papua New Guinea noting, for example, that in rural or remote areas they have few opportunities for employment and are unlikely to participate in 'stakeholder' meetings.

10   The sign Michael erected near the airstrip (Chapter 5) was certainly an act of overt resistance, in accusing outsiders of having destroyed a place sacred to his clan. Others in the community, however, paid little heed to the sign and openly disparaged the value of erecting it.

offended nor see that, at times, there was discord within the community. This latter concern was most strongly evident, publicly, when Amasai accompanied his classificatory sister to an aeroplane, chartered by Company, when she was en route to school. Under Company rules he had no right to approach the plane because he was not a passenger. The local security officer ordered him to leave the area. But that man had, in the past, assaulted the young woman. Amasai shouted at him: 'She is my sister now, she is not your sister'. He continued walking to the aeroplane. The altercation became more heated. Throughout the encounter, Amasai spoke in Tok Pisin; he wanted everyone, including outsiders, to hear and understand. Others in the crowd called him back but he ignored them. A local man, training as a loadmaster, spoke to him quietly and led him off the airstrip. The discomfort in the crowd was palpable. Amasai was publicising a local dispute. Matters such as this, people felt, should be hidden from Company representatives.

Much of what we have labelled 'arrogance' on the part of Company may be attributed to the perceived necessity of achieving required end points with minimal time delays. Consultation takes time and time is money. Company representatives often acted first and reacted later because they could, because they perceived themselves to be more knowledgeable than their hosts, because they perceived the task they were committed to undertake to be, ultimately, beneficial to all and sometimes because, despite the asserted 'cooperative' nature of lowlanders, they were anxious about security. Local people, however, from their own perspective, were also winners. By complying with Company's wishes and withholding evidence of internal discord, they ensured that Company remained at Suabi, that at least some monetary payments flowed to members of the community, that considerable savings of money or time were achieved when Company provided free helicopter or aeroplane trips for a number of school students and that they were less marginal than had formerly been the case. In achieving these outcomes, local people acceded to an understanding that they were the less powerful partner in the relationship. But they understood too that, as Sherry Ortner (1995: 175) expressed it, 'in a relationship of power, the dominant often has something to offer, and sometimes a great deal'. Though there were times when some people expressed dissatisfaction or indulged in minor acts of resistance, they did little that might jeopardise potential returns. Rather, they conformed to the collective desire for 'development' and, acting within the frame of conventional social practice, maintained an appearance of harmony and good relations.

What we saw as the 'arrogance' of Company and the 'complicity' of Kubo, then, are perhaps better recognised as epiphenomena. On both sides of the encounter the participants were being strategic and, at least in the short term, effective in achieving their objectives. Each set of participants was navigating the terrain of the other, though neither set fully understood that, ultimately, these were epistemological and ontological terrains that were given expression as moral imperatives. Company, in effect, treated local people as anonymous objects, to be manipulated and controlled for the greater good that would flow from success in its project. Kubo, in contrast, perceived Company representatives as subjects, to be engaged with in pursuit of productive relationships. But not all relationships are equivalent, and engagement may take different forms.

The Giving Environment

Nurit Bird-David (1990) depicted the economic systems of forest-dwelling Nayaka of southern India, Mbuti Pygmies of Zaire and Batek Negritos of Malaysia in terms of what she called 'the giving environment'. These people, she argued, perceive the forest environment as an 'ever-providing parent' that expects no return for what it gives. In a later article, reflecting on the post-contact activities of forest-dwellers, and with particular reference to Nayaka who supplement hunting and gathering with gardening, trade and wage labour, she argued that individuals or families often shift between different ways of appropriating resources (Bird-David 1994). Autonomy of action and an immediacy of response to changing circumstances are central to their mode of living.

The characteristics Bird-David outlined in these two articles are not confined to a particular set of hunter-gatherer societies. They appear as well in some Papua New Guinean societies that are better described as hunter-horticulturalists; in Sawiyanö of the Sepik (Guddemi 1992) and, we have argued, in Kubo (Dwyer and Minnegal 1998).

Kubo live in a world of uncertainty. It is relationships that matter here—relationships with others whose character is always contingent and can be known only through the ways that those others respond to one's actions. Social relationships are continually renegotiated; they can never be taken for granted. An ally one day may be an enemy the next, with sorcery an ever-present threat. Attention to others, and respect for others, is thus fundamental to sociality. Kubo valorise individual and family autonomy, but it is through sharing and reciprocity that commitment to relationships—mutual recognition of others as party to relationship—is revealed. And people pay close attention to the signs of mutual recognition that such exchanges constitute.

But, for Kubo, not all other beings are part of that social realm. The environment, too, is uncertain for Kubo—varying in unpredictable ways from day to day and year to year. It provides all that people need, in one form or another, but does not do so at the behest of people, or itself pay attention to the actions of people. Here, then, there is no need to 'give back', to reciprocate. This, to use Bird-David's language, is a 'giving environment'. People are free to take what they want or need from what happens to be available at any time. But this does not mean that the environment need not be treated with respect, that there is no need to pay attention to its vagaries. Kubo attend closely to the subtleties of what the environment is doing, not just so they can take full advantage of what it offers but also because there is always the possibility that the environment, or unseen forces that lurk within it, may withhold what is desired. The environment 'gives', but caution is advisable in 'taking'. There are forces at play that it is best not to antagonise.

At Gwaimasi, in the earliest years of our engagement with Kubo, people responded to opportunities and constraints as these appeared. Families or the entire community would, at times, switch from a predominant orientation to hunting-fishing-gathering-gardening to take up employment cutting survey lines or growing food on behalf of an exploration company. In the boom-bust environment that they experienced, this was an appropriate and life-affirming response. They were risk averse. As evidenced most strongly during the El Niño drought of 1997, they accommodated local uncertainty with respect to the availability of food by foregoing potentially higher returns in most years for the security afforded in years when food production was placed in jeopardy (Minnegal and Dwyer 2000a). To Kubo people, what is present—what the environment is affording now—is salient; it is the subject of both action and conversation. That which is currently absent lacks relevance. The environment is fickle. There is little point in being nostalgic about what is missing or what might have been. For example, during two weeks, late in 2011, people at Suabi harvested and ate vast quantities of the fruit of a species of *Pometia* (*kau ko*, a relative of lychee and rambutan). They collected the fruit after it had fallen into the river or after they felled a tree so that it dropped into the river. *Pometia* is a mast fruiter but our earlier visits to Kubo—five visits amounting to 25 months across 13 years—never coincided with the appearance of the fruit. Nor, in all those years did anyone mention it. Despite our interest in subsistence ecology, a focus of our earliest work, it was 25 years before we learned that *kau ko* was sometimes an important resource. To Kubo, environmental resources come and go. They are noteworthy only when they are present.

Shifting the predominant focus of interest was evident throughout our visits to Suabi in the years 2011 to 2014. In 2011, for example, people volunteered little about the fact that a company had been based on their land through the years 2006 to 2009. 'We ate rice every day' was the lingering memory though, in fact, there was much evidence of the past presence of that company in the form of material possessions, travel experiences and recent access to distant schools. And in that year, unless we prompted, no one referred to the PNG LNG Project. It was not active, and thus lacked salience. People's interests were redirected to the recent appearance of representatives of International Timbers and Stevedoring (IT&S) who, in exchange for lease-hold rights to land, promised a road that would reach from Kiunga to Port Moresby with side roads to and beyond Suabi. A year later, in 2012–13, few people talked of IT&S, for now attention was directed, initially, towards a forthcoming feast and, later, towards the arrival and potential promise of an exploration company. It was not until the latter half of 2013, when further exploration was undertaken at Juha and, more particularly, when the Department of Petroleum and Energy completed a clan-vetting process at Siabi, that people's attention was again directed to the PNG LNG Project and they became engrossed in the particularities of devising Incorporated Land Groups. Always, however, there were people or families whose focus was elsewhere: in 2011, for example, a Suabi family was based for several months at Famobi on the Strickland River, where they cared for their own and other people's pigs and planted and harvested a sweet potato garden on an island in the river. And, in 2014, there was a family who preferred life in the forest to life at Suabi and were away so often that initially they were forgotten by the man who, on our behalf, completed the census of Timaguibi Corner.

In the foregoing, our distinction between the 'social' and the 'environment' is both problematic and insufficient. For Kubo, there are many ways in which what we might see as natural is incorporated into the social and the cosmological (Dwyer 1996). A wild pig may be caught and domesticated, its potential for use in negotiating further relationships now predicated on the labour that its carer invests in it. A medium may marry a spirit woman, entering into a relationship of exchange with her kin. Invisible beings that inhabit the forest may be manifest in the bodies of animals, and intrude regularly upon the social life of people. The 'environment', as we have used it here, is a gloss. It refers to resources that are 'given', that require no, or minimal, input from people prior to their appropriation and which, on this count, for Kubo, are outside the social. In this sense, 'environment' is an emic category. What is included, and what is excluded, is a Kubo judgement—or, rather, our judgement of a Kubo judgement; a judgement, on our part, based on their practice.

Through the late 1980s and early 1990s, we observed Kubo people shifting their understanding of, and practice towards, 'foreigners'— missionaries, visiting Company representatives, government employees, anthropologists—from being potential relational others embedded in local social networks to being components of an environment that 'gave' without a requirement to reciprocate (Dwyer and Minnegal 1998). The 'foreigners' were often generous, sometimes frightening, but they lacked social commitment to Kubo. They did not participate in the necessary ethos of never-ending sharing. Like the environment that Kubo had always experienced, they were potentially beneficent but always unpredictable. In consequence, the conventional response to the locally persistent 'boom-bust' nature of access to environmental resources was extrapolated to encounters with, and expectations of, 'foreigners' who appeared intermittently but ultimately departed. It was the 'foreigners' who failed, who never took up the unexpressed but ever-present offer to become Kubo.

By 1995, Kubo were treating outsiders as components of an enlarged experience of environment. Outsiders held the potential to provide desired resources. All that was necessary was to find ways in which those resources could be garnered without risking refusal or worse. Avoiding the former, as we have described above, was facilitated by distancing oneself from a direct request. Avoiding the latter was trickier. A fish taken in the wrong place, a bird you have shot, a dog that bit you, a silhouette glimpsed in foliage, or words glimpsed on paper each had the potential to embody or attract the attention of a malevolent spirit being. The environment gave needed resources but was unpredictable. It could also give that which was not desired. Reciprocity was not necessary but the environment did not always comply with one's wishes.

By 2014 Kubo responses to outsiders had changed further. Our own position was ambiguous. To some people, we were social beings caught up in webs of friendship and reciprocal relations. We had, at the least, a taint of Kubo-ness that was given overt expression in the way that we lived. Our house was positioned within a cluster of their houses. It was small, with an enclosed and private sleeping area raised off the ground and an entirely open kitchen and work area on the ground. In many ways it was more like a traditional Kubo bush house than were any of the other sleeping houses at Suabi. It caught breezes and people enjoyed sleeping on the benches or watching while we typed or cooked. In our living arrangements, though never in our work, we were of the village. And not once in all those months were any of our belongings taken, though that would have been easy. We left much in the open. The front and back doors to our sleeping and storage area were of rough bush materials; it would have needed little imagination to gain entry. And to the surprise of many people we did our own cooking on an open fire. Indeed, one of the cooks from the Company camp filmed us as we cooked; he had never before seen white people cooking in this way. To many other people we were, perhaps, a harmless curiosity, nice to have around, tame 'whites'. Indeed, one of our friends often joked that we were 'albino Kubo'. But to most Suabi people, most of the time, we were as we had been from 1995 onwards—an asset, a means of acquiring resources and knowledge by those who learned best how these might be accessed. There were many days, sitting at the computer, typing yet another list of names, a business plan, a letter of request, or copying a reference that we hoped was not fraudulent, when we felt we had come to Suabi to act as village secretaries. There was, however, one significant difference. We had far less to offer by way of material goods, money or the future than did Company. It was with Company that people's response to the 'giving environment' had undergone most change.

The living arrangements of Company representatives were totally at odds with those of Kubo. There were hot showers, flush toilets, air-conditioned tents, refrigerators and freezers, electric stoves, covered walkways, computers and internet, satellite phones and ten or more containers holding surplus food, tools and equipment. The camp was self-contained. Staff from elsewhere seldom ventured out of sight of the camp and if the camp manager did go walking he was obliged, by Company regulations, to carry his radio and be accompanied by a security guard. In four and a half months there were only three occasions when a camp manager visited our house—800 m from the camp—to see, and be unimpressed by, the way in which we lived.[11] Only Suabi residents who were employed by Company were officially allowed to enter the camp and, though a few others did enter from time to time, it was usual that someone wanting to deliver a message or get their mobile phone charged would wait at a distance until they caught the attention of one of the local workers. Footwear was compulsory. In both its physical and social construction the camp was removed from the village and the people who lived there. It was, in this sense, dehumanised.

For many people at Suabi, there was an expectation that we would give, or that we had enough to give, without a requirement that we be reciprocated. But, always, it remained our prerogative as to whether we gave, as to whether we acknowledged and acted on our status as a component of environment. Company, however, unwittingly assumed a more distant and anonymous status. Some camp managers interacted warmly with the local men they employed and one, against Company expectations, sometimes swam in the river with children or went walking with a local man he had befriended. But Company seldom contributed to the daily social give and take of life at Suabi. Indeed, what we have reported as 'arrogance' on the part of Company was experienced by local people as indicative of Company's lack of participation in their own social world. For the most part, the representatives of Company did not pay attention to local people, did not 'respect' them, or recognise their social reality. It was, therefore, perceived as thoroughly legitimate to take from Company. Food stuffs, nails, tools and so forth were abundant,

---

11   We ourselves visited the Company camp on more than 20 occasions in 2013–14, both to introduce ourselves to the fly-in fly-out staff based there—including the five expatriate men who filled the role of camp manager through that period—and to offer or glean news. We were often invited to the mess for tea or coffee and sometimes enjoyed an evening meal. On one occasion, several senior Company representatives made a short visit to check the condition of the camp, fly over possible future exploration areas and move towards a decision about retaining or closing the camp. We visited to introduce ourselves, met the most senior of the visitors at the entrance to the camp and rapidly understood that he was not going to let us cross the threshold. We were not welcome. Our friends at Suabi were entertained by our dismissal and interpreted it as one more indication that we were Kubo-like. We note, however, that our engagement with Company employees was always as visitors and never as co-residents or co-workers. We were participants in several conversations in which white employees of one petroleum company disparaged purported approaches of another petroleum company or condemned the purported approaches of companies engaged in the timber industry. We recognise the diversity of backgrounds, knowledge and experiences of Company employees, and have observed differences in the ways individuals engage with and refer to local people, but have limited understanding of the ways in which these men engage with each other within or beyond the limits of the camp. Situated research is called for. In many ways, however, both in the field where critical balance is not always easy to achieve or subsequently when challenged by colleagues, this is fraught with difficulties (Coumans 2011; Golub 2014; Kirsch 2014; Welker 2016).

often unused, sometimes seemingly going to waste. They were there for the taking. Taking these things, without asking, was not 'resistance' any more than harvesting fish from a stream, or nuts from a tree, constituted resistance to the constraints of environment. Nor was it theft for, as church sermons sometimes made clear, stealing was sinful because it disturbed social relationships. Taking from either the forest or from Company was not disruptive in the same way. There was, however, always risk. The risk lay in overstepping the mark, in reaching beyond Company's threshold of tolerance—a threshold that could not be known in advance—and eliciting retaliation in the form of departure and loss of access to all that Company potentially offered. That is why people complied with Company expectations. That is why they were unwittingly complicit in the ontological persuasions of Company.

* * *

From the perspective of people at Suabi, Company was an objectified resource that could be freely exploited—with the proviso that those who represented Company were not made aware that, ultimately, this is what underlay local expressions of compliance. At the same time, however, throughout periods that Company was present, local people were negotiating with each other with respect to who would be employed, who had rights to land on which Company facilities were established and who, if anyone, had rights to the land on which Company would conduct their surveys. Here, then, irrespective of personal desires, their strategy was to conform to conventional expressions of morality and responsibility, by affirming reciprocal relations through ongoing exchanges and thus minimising the likelihood that they might be either subject to, or suspected of, sorcery. Stated simply, Company was not seen as party to the moral universe of people at Suabi.

Simultaneously, however, the 'complicity' of Suabi people—their behaviour as 'cooperative', welcoming, non-demanding and non-threatening hosts—reinforced a perception on the part of Company representatives that there was no need to attend to them, to negotiate relationships with them, as individuals. To Company, the people at Suabi were merely part of the equation of capitalist production—anonymous, substitutable, impersonal. Crucially, then, both the 'arrogance' of Company and the 'complicity' of Kubo were products of the 'friction' between two ontological systems, rather than being solely expressions of one or other of those systems.

8. The Things of the World

In mid-October 1991 we returned to Kubo-land after an absence of four years. The plane landed at Suabi and we shook hands and talked with many people who we had not seen for so long. One man told us that Bimo was at Gwaimasi waiting to see us. He had not come to meet us because he was not sure that we would return. Only when he saw us would he know we had come. But we did not know anyone named Bimo. 'You know him', the man said. 'He is Peter's friend.' It took some time before we understood that Bimo was the man we had known as Tobu; the man to whom Peter had been closest through our previous 15 months stay at Gwaimasi. But Tobu's name had changed. In 1986–87 his talent as a hunter was fading. He now seldom killed a pig. His returns from hunting trips were smaller fare: lizards, snakes and, particularly, birds. He became known as Bimo, for it was this species of pigeon that he was now most likely to shoot.

We walked to Gwaimasi and found few people present. As a government 'make-work' project, most were cutting a 'road' through swamp and forest to the mission station at Dahamo. And a recent death, attributed to sorcery, had seen people scatter to tiny forest shelters that were hidden and protected by fences from threatening forces. Bimo was working on the road. When we found him in the forest he was carrying Gabia, his three-year-old son who had not been born when we left at the close of 1987. Bimo made the introductions. 'My sister,' he told Gabia, indicating Monica. Peter—again, no name was mentioned—was introduced as Monica's *hwo de* ('good child'—husband). The implication was clear; Bimo had given his 'sister' as wife to Peter and thus these two men were deeply bonded as brothers-in-law. This meant as well, however, that Peter was in debt, for he had not yet reciprocated Bimo's 'gift'. Bimo's form of introduction was an invitation—neither a request nor a demand—to accept the relationships he had mapped out. It was up to us to respond through time as we chose.

Through the preceding chapters we have traced various shifts in the ways that Kubo people interacted, and sought to manipulate relationships, not only with each other but also with the new beings, things and powers that have appeared in their lives through recent decades. But lurking behind these changes have been deeper shifts in modes of understanding. In this chapter we bring together implications of the diverse accounts presented earlier. We do so to highlight the interplay of changing epistemologies and ontologies in the strategies people deployed to navigate both future and past. The ways in which Kubo and Febi people come to be known and named have changed through the time we have lived and worked with them. Those changes reflect deeper shifts in how relationships are understood and negotiated.

Naming Persons

When we first lived with Kubo people at Gwaimasi, in 1986–87, children were not named until about the time that they walked. Until then they were spoken of or addressed as *sobosio* (girl/woman-like) or *oosisio* (boy/man-like). The implication was clear; they may have had the superficial appearance of being male or female, but even this basic aspect of identity remained uncertain. Indeed Gehogwa, who had attended school for five years, translated these terms as, respectively, 'false girl' and 'false boy'. The uncertainty remained until a child began to reveal, through its actions, who he or she was. Only then was the child given a personal name—an appropriate 'custom name' that invoked connections to past people, places or events. A particular person might be given the right to name the child, and thus position the new person in relation to themselves and their world, to reproduce the past in the present: the name of an ancestor, perhaps, who had shaped the social and physical world; the name of someone close who had recently died—the namer's father, perhaps, or a sister; or the name of a place, or of a being that inhabits that place, interaction with which is now less common. Even then, however, there might be ambiguity, with several names tested before consensus about the child's identity was reached. One small girl, for example, moved from being Sobosio to Bosai (like Sai) and, only later, as her similarity to the long-deceased Sai seemed to be confirmed in her behaviour, did her name become Sai. These 'custom' names were powerful, evoking the larger social domain from which they have been drawn, and people used them with care. We never learned the custom names of some people at Gwaimasi.

Other names emerged as the growing child interacted with the world in the present; names that reflected, perhaps, something of the person's appearance—'grass grub', or 'tall man'. Or the names reflected actions and behaviour—Hwo ton (dead child), for a boy who had suffered a fit as an infant and was thought to have died, or Bimo for someone whose hunting now focuses on procuring fruit doves rather than pigs or fish. These names were less potent than 'custom' names, but were still used with care; people preferentially referred to others by kin terms. And as interactions played out, names might change, new names become recognised as more appropriate; the boy known first as Hwo ton later became Gawua (grass grub) and eventually Fuhuwa. Such names were always contingent, grounded in the present of everyday life.

If, however, a child was born at a community health centre, or had been otherwise hospitalised or vaccinated, then medical staff—if they were not of the home community—required, and often themselves suggested, a name for their records.[1] These names were usually taken from English or from the Bible—though some mothers cleverly suggested variants on 'Baby' (for example, Bebi)—and gained currency before a custom name was discovered.

Custom names, and the other names discussed above, emerged out of the contingencies of everyday life. They were found to be appropriate by other people and attached to a person, who, at that time, could never say their own name. But, again, at that time, and increasingly through the years that followed, some people—particularly youths or young men—chose an English name for themselves. These modern names were an expression, by the name's bearer, of who he or she would like to be, or be like; unlike custom names they sought to shape the future, not merely reproduce the past. These names, increasingly, are drawn from the outside world. Some are from the Bible—Elijah or Israel, or Peter and Paul and Rebecca. Others have more secular connotations. Jackson was a common choice a decade ago. One young man chose Elton, and is also known as John. Betty, Diana and Sandra all live at Suabi. And, finally, there were other names—reciprocal names—that two people might share to mark an incident or occasion they had had enjoyed together. At Gwaimasi there were two men who addressed each other as Gwamo Dihio (frog eye) in

---

1 At Suabi, in 2011–14, the community health worker was a local man. His records of births include the name of the mother and, if known, father but do not provide a name for the child.

memory of a time when they had carefully shared a small frog—they had each eaten an eye—which one of them had captured; no one else addressed either of them with this name.[2]

By 2011–14, at Suabi, all the above kinds of names continued to have currency but there had been shifts in the time at which they were first used and in the frequency of different kinds of names. Now, very soon after birth, a child would receive either one or two names: a custom name and an English or biblical name. If they received both, the custom name might initially be withheld from public knowledge, to allow time to learn whether it was indeed appropriate. In these cases, then, it was the English or biblical name that rapidly gained currency. We did not hear infants addressed as *sobosio* or *oosisio*. As before, custom names could alter, or shift back and forth, through the course of a person's life but, to a much greater extent than earlier, young men and women chose for themselves an English name and, with varying success, encouraged people to use that in both address and reference. They were now far less reticent to speak their own name. Indeed, many earlier name taboos had been abandoned though most people were hesitant, especially in public, to use the name of an affine and were very resistant to naming the clan of their mother-in-law.

In 1986–87 and, indeed, through the next decade, people seldom used 'surnames' that directed attention to a relationship (with parents) that they had played no part in producing. This practice had been introduced by patrol officers when they took a village census; it was usual to ask for the father's name and record this as an analogue to a surname. And, later, these double-barrelled names were used by government officials who issued documentation that a person was eligible to vote. At Gwaimasi, in 1987, when an election was pending, we were brought a package containing the electoral papers of people who were registered to vote. These papers had been stored together for safe-keeping, but no one could read and so they could not discern which document belonged to which person. We were asked, therefore, to distribute the papers. But there was a problem; we did not recognise any of the recorded names. No one had stated their own name when registering—to speak it was taboo—and

---

2   Knauft (2016: 14) termed these 'gift exchange names', and reported that among Gebusi such names link each pair of men in a village, with this being the most common mode of address and even of reference. At Gwaimasi, such names were much less common; certainly, not all pairs of men used reciprocal names.

the names that had been offered as father's names were seldom those of the person's biological father. We read out the names, and people stepped forward to claim their registration paper.

Gradually, however, double-barrelled names assumed increasing importance. They were required for school enrolment lists, and as people began to write letters—especially official letters to exploration companies or government departments—became adopted as standard practice. It was most common that father's name would be used but, in some circumstances, people used the name of the male 'head' of the house in which they lived. Some young men—particularly those who had no biological sister—chose to use the name of a different, senior male kinsman as their father's name in an attempt to enhance their options with respect to marriage partners. Similarly, where a man had more sons than daughters, a girl from a related family might be 'adopted' as an exchange sibling for one of the boys and, from then on, name that man as her father. And, not infrequently, where a man had been killed or diagnosed as a sorcerer his child was ashamed and chose the name of a father's 'brother' to use as father's name.

Among Kubo and Febi, as our opening example made clear, customary naming practices were contingent and relational. Indeed, the relationship was of more significance than the name itself. The names emerged in the contexts of a person's perceived actions and appearances and, as these altered through time, so their names might alter. In 1986–87, at Gwaimasi, we learned that the father of the boy Sigio was Tameho. There was no one at the village with this name, and Sigio associated with older classificatory brothers. We assumed that Sigio's father had died. But, after some months, the married couple Fafobia and Uhabo argued and, for a few weeks, Fafobia moved elsewhere. Sigio immediately realigned with Uhabo and, by everyone at Gwaimasi, Uhabo was immediately spoken of, and to, as Tameho. Clearly, the connotations of those names—the person each name evoked—were different; Uhabo was Fafobia's husband but Tameho was Sigio's father. The relationship that prevailed at any time elicited the appropriate name. Indeed, when Fafobia returned to the village, Sigio moved out of her house and his father once again became Uhabo.

The name that had currency at any time, then, marked the relationship between the bearer of that name and those with whom he or she interacted. Names emerged in place as a reflection of who, or what, that person was or had become. They emerged in the context of community and, indeed, given that their use was taboo to the name's bearer, he or she had minimal influence over their form. They expressed a collective judgement about salient relationships and, though for some people the one name might persist for decades, there was never any certainty that a person would not change and reveal a new name. Indeed, a common expression of change occurred at the time a child was first given a name—as Monica, for example—and its parents from then on became known, in address and reference, by the teknonyms Monica-*dua* (-mother) or Monica-*ade* (-father).[3] Here, again, therefore, relational configurations were prioritised in naming.

In several ways, the changes in naming practices since colonisation have reduced the importance of the relational. This is evident, first, in that naming a child very early in life reduces opportunities for that child, through its actions, to reveal an appropriate name to the people with whom it increasingly interacts. It is evident, even more forcefully, when individuals choose their own name and, with varying success—some older women, in particular, are resistant—encourage others to adopt it. It is evident too in the fact that so many people are now willing to speak and, of course, write their own name and, finally, in the 'fixity' that arises from the perceived need to satisfy outsiders—government and Company—by consistent use of the one double-barrelled name.[4] Where, once, names reflected who or what a person was or had become at a particular time and in a particular context, names at Suabi were now, increasingly, pre-emptive. Increasingly, names positioned a person; the name itself, and not ever-shifting actions and appearances, now made a person what or who he or she was. Whereas, in the past, it might be said that a person's interactions revealed their name, it is now much more the case that one's name is presumed to reveal one's personality (compare Harrison 1990: 59). And as names come to be written down—on birth registers or in vaccination records, on census or election rolls, on wage sheets or references—names

---

3   'Monica-*dua*' does not mean 'mother of Monica' which would be expressed as 'Monica *ba dua*'. Teknoymns of this sort were not common in 2011–14, though self-reciprocal kinship terms retained currency and were sometimes expressed in English. Thus, for example, *ade* is the Kubo reciprocal term of address between father and child (Minnegal and Dwyer 1999: 65–6) but, at Suabi, a father would sometimes address his young son as 'Daddy'. Similarly, 'uncle' often replaced the reciprocal term '*babo*' when a man addressed his sister's son.

4   Harry Walker (2013: 139–40) discusses the adoption of Spanish names by Amazonian Urarina people and the ways in which these are employed to confer legitimacy within contexts of the Peruvian state.

become immutable, carried between contexts. Thus the way in which people were known, or knew themselves, was shifting profoundly. In the domain of naming, as in so many other domains, fluidity was giving way to fixity as a predominantly relational epistemology was progressively eroded.

Naming Groups

The erosion of a relational epistemology was even more evident in the ways in which sets of people were known, named and placed on the ground. By 2011–14, households had assumed a salience that had not, previously, been the case. Family had always been important; a husband, wife and their young children undertook many subsistence tasks as a unit—family autonomy was valued highly (Dwyer and Minnegal 1998: 29–30)—but, progressively, the spatial separation from others that this entailed carried over to periods people spent at the village itself. The shift had commenced by 1986, initially with a move from communal longhouses to villages as clusters of houses in which, at first, two closely related families—exchange brothers-in-law, for example—might share a house and, later, in conformity with an emerging expectation that each married man with his wife, or wives, and children should live separately in their own house.[5] Particular men became recognised, and named on documents, as 'head of household' and each 'household' stood alone.

While households were increasingly recognised as distinct, the layout of villages strongly reflected relationships between them. At Gwaimasi, through the 1980s and 1990s, the focal relationships were affinal; men who had exchanged sisters tended to build adjacent houses, with men of different *oobi* thus interspersed through the village (Minnegal 1994: 82). At Suabi too, in 2013, it was not uncommon for a man to build a house near that of his wife's family. But this was changing. In January 2014, at the market and at other informal meetings, there was much discussion about the need to clean the village Corners, improve hygiene by relocating toilets, confine pigs to fenced enclosures and alter layout from a perceived haphazardness to a more orderly arrangement with central

---

5   At Suabi, in 2011–14, one man had built separate houses for each of his two wives and the children born to each of them.

streets and individually fenced houses.[6] Some men argued forcefully that families whose 'household heads' were members of the same clan should establish houses as a single cluster, though others suggested that an arrangement of this sort was more likely to promote, rather than alleviate, disputes. Several work parties formed to clean village Corners, and a burst of building activity commenced, as men built new houses near those of their fathers and brothers. While some of the more ambitious proposals were not taken up, all reflected a perception of order that was structured according to defined and bounded assemblages of people. Now, however, the assemblages were to be defined by agnatic relationships between men, rather than those established through marriage or initiation.

More striking, and of greater significance, was the reordering and restructuring implicit at the level of 'clans' that occurred through time and informed the vision of a new village structure. In 1986–87 neither the English term 'clan' nor the Tok Pisin '*klen*' was used by Kubo people. It was *oobi* identification that was salient and in our own earlier writings we, too, hastily, glossed *oobi* as 'clan' (Minnegal and Dwyer 1999, 2011a). There was, in those earlier years, relatively little ambiguity about *oobi* affiliation. Though people would, if asked (usually by us), readily name the *oobi* with which they identified, there were few occasions when *oobi* identity featured in the course of spontaneous conversation. Admittedly, when marriages occurred—when new relationships were being negotiated— the *oobi* affiliations of the partners was a topic of keen interest and, if the liaison was judged inappropriate, of gossip. But it was what particular people were doing, where they were doing it, with whom they were doing it and, always, what they were eating that was of abiding interest and the subject of daily conversation.

Oobi identification, then, had its place but was of little relevance to people's daily lives. Indeed, as noted in Chapter 2, it was day-to-day practice that informed *oobi* identity. A child raised by a man from a different *oobi* might well identify, and be identified by others, as associated with the *oobi* of his stepfather, not his biological father. And the association of people from a given *oobi* with a particular area of land was similarly mutable. No one spoke of 'borders' that marked the land of one *oobi* off from that of another. Rather, people from adjoining *oobi*

---

6   One advocate of these suggestions—a former Suabi community health worker—drew on the Healthy Islands Concept that he had encountered in a recent course on health education and which has been promoted elsewhere in PNG (Temu and Chen 1999).

were 'brothers', they should not intermarry and their lands, activities and patterns of residence flowed diffusely one into the other. On the ground, *oobi* were fuzzy (compare Hays 1993)—they lacked definition; the social relations they inscribed achieved certainty only with distance (Minnegal and Dwyer 1999: 68–9).

By 2011–14 people at Suabi spoke of 'clans'—the term *oobi* was seldom used—and, as described in earlier chapters, these were now conceptualised, and to some degree documented, as coherent sets of known people who were associated with a known, and bounded, area of land. Certainly, in practice, neither the sets of people nor the place had been irrevocably decided, let alone accepted by government.[7] But the ideal and intention were there. Clans and their subdivisions—subclans—were conceptually fixed, their expression infiltrated, it often seemed, by earlier and mistaken models drawn from the anthropology of New Guinea (compare Golub 2007: 39–40). The task for many people was to decide who to include and who to leave out. The task for many others was to have a desired association recognised and ratified on paper. In attending to these tasks a relational epistemology was to the fore. People called on both past and present connections—mythological and mundane—to assert or deny rights. But at a higher level these details were embedded within, and subservient to, a categorical configuration of social order; a hierarchy of

---

7   In 2011–14 most lists of members of Incorporated Land Groups (ILGs) nominated the 'block'—the graticular block—with which that ILG was asserted to be associated and with which, therefore, the members of the ILG were asserted to hold rights. The blocks were identified by their official number—assigned by the state—and by a name chosen by local people. Two blocks—named as Siagu and Tihin—were nominated most often; three others—Bebesy, Bogubi and Wasiga—were nominated once only. Siagu is the name of a place in the southeast of Febi territory. It is possible that Wasiga is the Febi name for a species of tree kangaroo; we had earlier recorded Wasigia (*will trick*) as a Kubo name for a species of mammal, noting the likelihood that it was a tree kangaroo. We have no information on possible sources of the other names given to these blocks. Inherent in this naming practice, however, is the first indication that land could be marked out, and ultimately conceptualised, according to the dictates of an abstract geometry. But, while Kubo and Febi people were reframing geographical understandings in terms of abstract borders—of lines on maps—some Westerners who visited their lands were, for quite different reasons, losing any sense of local position in space. During four-and-a-half months at Suabi in 2013–14, five men from Australia and New Zealand served as camp managers at the exploration camp. They were 'fly-in, fly-out' workers, arriving by plane or helicopter from Mount Hagen, staying for a few weeks and, with one exception, not moving even as much as one kilometre from the camp during that time. None of these five men knew where he was relative to the Strickland River; that is, none knew whether Suabi was east or west of that river. The circumstances in which they worked, with excellent radio and satellite phone facilities, provided the guarantee that, should the need arise, they could be rapidly evacuated. They had no need for local geographical knowledge. It had become irrelevant to the way in which they lived.

clans and subclans each of which was divided according to the status and attributed rights of acknowledged members, whoever they might prove to be.

The clans and subclans that Kubo and Febi people spoke of and wrote about in 2014 were increasingly, in their imagination at least, individuated (Minnegal and Dwyer 1998) or, as Thomas Ernst (1999) expressed it, 'entified'. They stood apart from other clans and subclans. They were not as they had been before, and people's affiliations with them, likewise, were not as they had been before. Clans were emerging as bounded, as individuated. And, increasingly, persons were emerging as individuals, defined by the attributes they possessed rather than the relationships established with others. The world of 'things' that lent order to the lives of Kubo and Febi people was changing. In the imagination of these people new, different, social 'things' were coming into existence and were being ordered in new ways. The ontological framing of their world was in flux.

The Known and the Knowing

Things are both necessary and inevitable. This is because the inherent ambiguity of relations in complex, open and autopoietic systems—systems of life and imagination—can be only obviated where 'parts' of the system appear as 'individuated' (Dwyer 2005: 20, n. 12). It is at the boundaries between those 'parts' that exchange occurs, that 'noise' is reduced, communication is facilitated and, in human systems, agency may be expressed. Those boundaries may be real or imagined: 'they are created, repositioned, transformed and sometimes extinguished as systems change' (Minnegal and Dwyer 2011a: 316). Boundaries give expression to concatenations of relations that may be perceived and conceptualised as 'things'. It is often the case that names act in this way in appearing to specify a domain of relations.

Anthropology has a renewed interest in material objects as 'things', showing how they may be deeply enmeshed in social and mental life (Appadurai 1986; Bell 2009; Malafouris and Renfrew 2010; Zeitlyn and Just 2014: 8–9) and often arguing, though this is not our position, that they have agency in their own right (Gell 1998).[8] But, as Benedict Anderson (1991) made clear in *Imagined Communities*, 'things', as understood, acted on and acted with, by people need not have material expression. They may be figments of the imagination, though their effects may still be profound. Anderson's concern was with the nation as an imagined political community. He attributed its emergence to the development of printing in vernacular languages and to the ways in which this facilitated a common understanding among groups of people who might never encounter one another. The lists of members of Incorporated Land Groups (ILGs) drawn up by Kubo and Febi people—lists that profess a corporate identity for a bounded set of named people, at least some of whom may not know each other—might be similarly appreciated as 'imagined communities' in the making.

Ontology concerns the 'things' of the world, what they are and how they are ordered. But what things are understood to be, and how they are ordered, need not be the same for all people or for the same person through time. There was a time, for example, when ultimate physical reality could be reduced to atoms, then later to protons, neutrons and electrons, and now, it seems, to a particle—the Higgs boson—that gives mass to other particles. This in turn, perhaps to come, may lead to yet other refinements of the ontological universe in which some people dwell. There is, as yet, no theory of everything.

Ontology is not static. It shifts as people experience, metaphorise, and ultimately concretise, the world in different ways. There is no unchanging 'essence' or 'Being'. Nor, it must be stressed, does that which is known to people—the 'things' of their world—necessarily assume priority in their daily lives. The 'things' they come to know are epiphenomenal

---

8   We take agency to refer to acting on the world rather than merely acting in the world. It is 'intentional causal intervention in the world' (Ratner 2000: 413), though that neither precludes a plethora of constraints on action nor implies that 'intentions' necessarily come to pass (Dwyer and Minnegal 2007). With respect to the outcomes of agency, contingency is unavoidable. We accept that there are people, including Kubo and Febi, who consider that some species other than human, and some material objects, have a capacity for 'intentional causal intervention in the world' or may harbour beings that have this capacity. They understand agency in much the same way as we do but attribute this capacity to a wider range of things (compare Povinelli 1995; Morphy 2009). Some of them appreciate this difference, as was made clear when, at Suabi, one man observed that 'white men' do not think that snakes have 'life'. We had been asked how we would react if one of us was bitten by a death adder and died. Our initial response, that the one who had not died would be sad but would not seek redress, was dismissed; it took no account of the fact that, to Kubo people, death could happen only if the death adder had been empowered to act on the world—and, thus, had 'life' (agency)—by a malevolent sorcerer.

concatenations of relations—they appear as 'entities'—but, to some people, in some circumstances, the relationships between those 'things' are more fundamental than the 'things' themselves (Wildman 2010: 54; compare Wagner 1977). This, Wesley Wildman (2010: 54) wrote, qualifies as a relational ontology, to be contrasted with a substantivist ontology in which the 'things' are primary and the relations derivative. The distinction between the dividual and the individual person provides an example from anthropology, with the former relational and the latter substantivist (Strathern 1988; Busby 1997; Mosko 2010). Brent Slife (2004: 159) wrote similarly, contrasting a relational ontology in which 'all things … have a shared being and a mutual constitution' that is immanent in practice, with 'abstractionism' in which 'theories, techniques, and principles, capture and embody the fundamentally real' but may be distanced from day-to-day practice (ibid.: 157). Analytical dichotomies are, of course, always risky—too easily misread as rendering that which is analogical and messy as digital and unambiguous—but our take on the contrast to which Wildman and Slife draw attention is to think in terms of relational and categorical ontologies as co-existing potentialities. Or, better, to acknowledge that there are domains of people's imaginary and quotidian lives in which their experience of 'things' may prioritise either relations or categories.

The shift from relational to categorical ontology that we envisage emerges in practice and does so via a process of abstraction—a distancing from lived experience that, simultaneously, 'transforms that experience and creates contexts for further transformation' (Dwyer and Minnegal 2014: 51). There should be no expectation, therefore, that the lifeway of a particular people—some philosophers excepted—at a particular time will be underwritten by a coherent ontology (Keane 2013: 189). We should expect, rather, that in some domains of people's lives the 'things' that fall within that domain are experienced, understood and employed by reference to the relations they have with other 'things' while, in other domains, 'things' are experienced, understood and employed as standing alone, as concrete 'objects' (compare Ingold 2010: 4).

But here, of course, we have intruded on the epistemological, on the ways in which people know the world of 'things' within which they dwell. If ontology can be represented, for a particular people, by reference to their experience and understanding of 'things' in their world then, while allowing that the 'things' can change, epistemology can be represented by reference to their understanding and employment of those 'things' and the truths and beliefs that, for them, thereby, follow. And, once again, this is manifest in practice. As Mario Blaser (2009a: 877) wrote: 'ontologies do not precede mundane practices, rather [they] are shaped through the practices and interactions of both human and non-humans'. The same may be said of epistemologies.

Again, as with ontologies, and as argued throughout this book, epistemologies may be appreciated as either relational or categorical and, again, as with ontologies, it will be in particular domains of experienced worlds, and not the entirety, that relational or categorical ways of knowing are prioritised. Nurit Bird-David (1999, 2006), writing of animism and the hunter-gathering Nayaka of southern India, states the matter clearly. A relational epistemology, she wrote, is 'about knowing the world by focusing primarily on relatednesses, from a related point of view, within the shifting horizons of the related viewer. The knowing grows from and *is* the knower's skills of maintaining relatedness with the known' (1999: S69).

Her emphasis, as is ours, is with practice—with the phenomenological. She contrasted a relational epistemology with a modernist epistemology, declaring the latter to comprise 'a totalising scheme of separated essences, approached ideally from a separated viewpoint' (1999: S77) and asserted that: 'Framing the environment relationally does not constitute Nayaka's only way of knowing their environment, though in my understanding they regard it as authoritative among their other ways' (1999: S78).

There are contexts, Bird-David allowed, in which modernist—we would say categorical—ways of knowing infiltrate Nayaka epistemology, and in a later article, writing of change and of shifts to cultivation and animal husbandry, she demonstrated this with respect to contexts in which some forest and domesticated animals or plants were treated as objects (Naveh and Bird-David 2014).

To capture the difference between relational and modernist epistemologies, Bird-David contrasts the way in which, on the one hand, Nayaka 'live' with the forest and, on the other, botanists 'study' the forest. The former are attentive 'to variances and invariances in behavior and response of things in states of relatedness and for getting to know such things as they change through the vicissitudes over time of the engagement with them', they are attentive to what a tree does 'as one acts towards it, being aware concurrently of changes in oneself and the tree' such that they grow 'into mutual responsiveness and, furthermore, possibly into mutual responsibility' (Bird-David 1999: S77). The botanists, by contrast, acquire knowledge 'through the separation of knower and known and, often, furthermore, by breaking the known down into its parts in order to know it' (ibid.; see also Wagner 1977, Gow 1995 and Scott 1996 for comparisons of knowledge construction by scientists, on the one hand, and, respectively, indigenous Papuans, Amazonians and north Americans).

Among Kubo and Febi the way in which children were, and are now, named may be understood in a similar fashion. In customary practice, a child's name emerged in the context of his or her interactions with others. The name was revealed in practice, in the ongoing and ever-shifting 'mutual responsiveness' between the growing child and the community of others who came to know the child. The name connoted those relationships, and through time, as that child developed and his or her actions and appearances altered—as he or she became other than they had been and, hence, related in other ways to other people—so too his or her name might alter. Now, however, a child's name is attached in advance of any experienced 'knowing' what, or who, that child is becoming. The name no longer emerges through the child's relationships with others but, rather, from the knowledge—a new kind of knowledge—that this child stands apart from all that impinges upon him or her.

But this shift in naming practice was not absolute. In 2011, it was the 'chosen' names that many people at Suabi favoured, and preferred that we and others use. By 2014, however, nicknames or custom names were again more commonly used to identify people in both address and reference. Kubo were reorienting, it seemed, to relationships of the past and the present in marking identities. This does not mean that the future had become less salient. Rather, people were now looking to the past—reconfiguring the past—as a way to position themselves for what they envisaged would be a radically different future. Being able to take full advantage of new possibilities that might open up required that identities be securely grounded both in relationships of and with the past and in interactions in the present. Which name was prioritised depended on context. Similarly, in 2014, in the context of feasts Suabi residents foregrounded the relational concomitants of food—of, for example, pigs or pots of rice—in distributions, while in the context of exchanging a pig as compensation or bridewealth 'a pig was just a pig'; its categorical concomitants were foregrounded. And, again, the ILG lists produced through 2014 included people on the basis of productive relationships with the inscriber, but were to be presented to government as specifying the bounded set of those who shared 'membership' in, and thus rights to the land of, a named 'subclan'.

Finally, as Bird-David (2006: 43) insisted, 'knowings are always and everywhere inseparable from the knowns' (see also Dewey and Bentley 1975; Ingold 2000: 111–2). She argues that the 'animistic ontology' of Nayaka is 'inseparable from their animistic epistemology' (Bird-David 2006: 34). Ontologies and epistemologies travel through time as a package. A shift from relational to categorical in the former will establish the ground upon which there will be a coincident shift from relational to categorical in the latter.

Global Flows

Anna Tsing (2009, 2013) has shown how some local resources, matsutake mushrooms for example, may enter extensive supply chains, filtering via local and itinerant harvesters and a succession of buyers, distributors, processors, exporters, importers and retailers—gaining value at each step—to eventually be sold at high prices to elite international consumers. She has shown, too, how these supply chains irrevocably entangle biology, environment, economics, politics and multiple expressions of social life in ways such that each of these domains 'rubs up' against the others to ultimately create capitalist commodities from what were once 'free to take' natural resources. It is in the interplay of these domains, in their never-ending engagement one with another, that people act, change is facilitated and a semblance of, admittedly, heterogeneous order is generated. It is that interplay—that rubbing up against each other—that Tsing (2004) has metaphorised as 'friction'.

Commodity chains are usually discussed as though they are unidirectional, with the direction being from local to global. As Tsing makes clear, this need not imply that those at the origin point consider themselves oppressed. Indeed, the mushroom hunters she worked with in Oregon regard their activities as an expression of 'freedom' and the mushrooms they gather as embodying social relations and, therefore, as 'gifts' (2013: 26–32). But though they may understand themselves to be free, the system in which they are entangled progressively erodes social relations as those mushrooms move along the chain. Their gifts, as Tsing argues, become commodities. To the elite consumers the mushroom hunters are anonymous, encompassed by a system of interconnections and frictions that the latter embrace but that, ultimately, depersonalises them. The morality of the gift—a morality in which they are beholden to known relational others—is rendered subordinate to a morality in which they are beholden to anonymous and controlling outsiders (compare Minnegal and Dwyer 2011b: 205–9). Or, as Tsing wrote: 'The new subjects of liberalism are even more trapped in power because they imagine it as freedom' (2004: 215). Ultimately, there are no social relations that connect persons at the two ends of the chain.

In fact, however, the flow is never unidirectional (Appadurai 1990; Bestor 2001). This is made clear in *Ethnicity Inc.* in which John and Jean Comaroff (2009) show how, among others, native Americans, Shipibo of Peru, Catalonians of Spain and Agukuku of Kenya have appropriated some of the trappings of modernity—particularly those of the market place—to forge both viable commercial ventures (casinos, tourism) and new, revitalised identities. Indeed, often, it is identity itself that is marketed and, 'with a good measure of critical and tactical consciousness', corporatised (ibid.: 27). At one level, then, this is the 'invention of tradition' writ large (Hobsbawm and Ranger 1983; Jolly and Thomas 1992) and, though it risks reifying 'culture', it 'also has an impact on everyday conduct: on those less-objectified, unremarked upon ways of doing things—even things instrumental, bureaucratic, and commercial— that embed themselves, "thickly" or "thinly," in local conventions, styles, and values' (Comaroff and Comaroff 2009: 75).

But what is that process and what is it that, at base, is being appropriated? The ways in which people change how they dress, talk, organise spaces in which they live, selectively acquire material possessions, educate their children, alter religious practices and accommodate to a monetary economy are likely to be immediately apparent. But these are surface expressions only. The process itself runs deeper. It entails ideas taking on a life of their own, values being measured against a common standard, the attributes of things assuming definitional status, and transactions foregoing prior relational commitment (Minnegal and Dwyer 2011b: 208). These are the neoliberal persuasions of, respectively, reification, commensurability, categorisation and anonymisation. All are party to commodification. It is these expressions of process, and their attendant tropes, that flow from the global to the local and, in the ways they infiltrate and inform local practice and understandings, underlie social transformation (Minnegal and Dwyer 2007). The multiple encounters entailed in commodity chains are those of sometimes subtly different, sometimes profoundly different, knowledge systems; they are ontological and epistemological encounters.

Kubo and Febi people living at and near Suabi are not harvesting natural resources that then enter a supply chain of the sort Tsing and others have described, though in the recent past a few have sold small quantities of agarwood to buyers, and there are some who contemplate future possibilities with this highly valued resource. Nor, of course, have these people developed tourist ventures, let alone casinos, though, again, some know that elsewhere in Papua New Guinea (PNG) there are people who derive income from marketing 'traditional' ceremonies or being engaged in eco-tourism (Errington and Gewertz 1996; Otto and Verloop 1996; Silverman 1999; MacCarthy 2013; Sakata and Prideaux 2013). What they have done, however, is to provide labour and services—use of land, access to water and so forth—to mining and oil and gas companies that have explored, and established infrastructure, on or near their lands. For the companies concerned, that labour and those services are cheap. In late 2014 and early 2015 oil and natural gas prices collapsed worldwide and many companies put new ventures on hold and reduced engagement with existing ventures. But this was not the response in PNG where, for example, the managing director of Oil Search said that the prospects for Liquefied Natural Gas (LNG) expansion remained 'attractive based on the current oil price outlook' (Macdonald-Smith 2015). The cheap labour and services provided by local people contribute substantially to the economic viability of LNG projects in that country.

The labour and services provided by Kubo and Febi people are, in fact, the origin point of a different kind of supply chain. The links are again numerous. The chain reaches from local participants as workers and hosts, through the field officers of companies and a hierarchy of officials of those companies, to the stock market and, ultimately, to shareholders. And, again, of course, to those shareholders the people at the origin point of the chain are anonymous and non-relational. For Kubo and Febi people, however, the immediate source of new experience and new knowledge is the presence, and the activities, of those officers of companies who visit and live on their lands. It is through their immediate engagement, in time and place, with those who come to their lands that they encounter different ways of living in, and understanding, the world and that their pre-existing ontology and epistemology are challenged and may be transformed. How then might we characterise the worlds of, on the one hand, Kubo and Febi people and, on the other, Company representatives when they meet on the lands of the former people? How are their different practices framed? It is in the differences between the two that new ways of living emerge and give shape to new 'things' and new ways of knowing those 'things'.

At the Company Camp

The epistemological, ontological and moral underpinnings of Company, and by extension of Company's representatives, differed from those of Kubo and Febi people. The abstract entity that is Company occupies a world of risk, not uncertainty. The focal environment is not experienced as 'giving'; the gas being sought is already there and once found in sufficient quantity may be extracted. Market considerations, and the potential rewards from resource exploration, favour a risk-prone strategy under which the adverse outcomes of some ventures are expected to be more than balanced by the high returns of others. Gas may, or may not, be present in any place, and the value—the viability—of a find may fluctuate with the market. But probabilities can be calculated before action is taken. Inasmuch as considerations of science and economics—of technology and the market—lie at the core of practice, so too, for Company and its representatives 'on the ground', both the social world and the material world are grasped and experienced largely in categorical terms.

The camp that Company established at Suabi in 2013–14 was ordered with respect to both structure and performance. There were street signs and arrows to direct people to work areas, sleeping quarters, mess halls, clinic, ablution blocks, a muster point where people should gather in emergencies and so forth (Fig. 8.1). A notice board listed the names, responsibilities, company of origin and allocated accommodation of senior employees. Another, open to the public, showed the campsite fire plan, reported weekly statistics about man-hours worked and safety incidents, and provided information about what to do in the event of heat exhaustion, excess noise exposure and medical emergencies. Covered walkways, of timber and wire mesh to minimise slipping, connected different sections of the site. The site was bounded, an island of difference from the greater village of Suabi within which it was located. It was a place of and for people who were themselves spatially bounded by the structures that enclosed them, and who therefore, on this count alone, were different to the people on whose land they now lived.

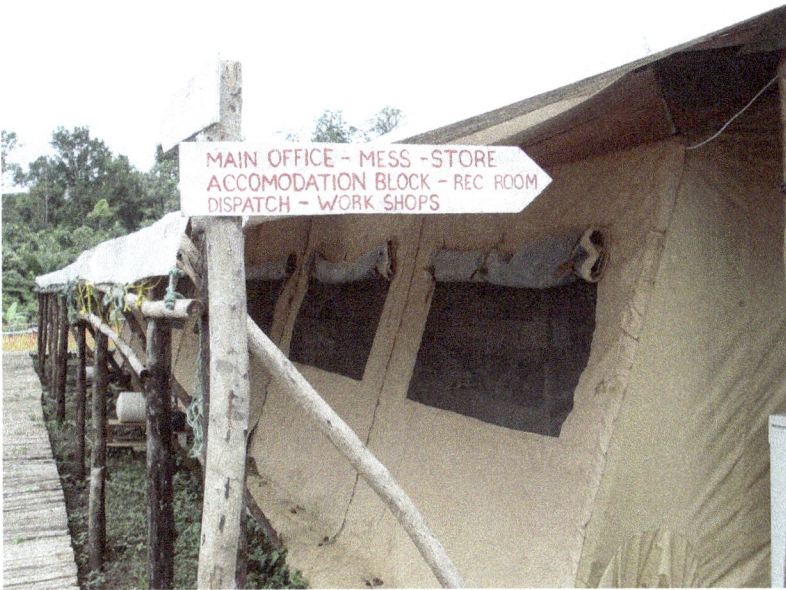

Figure 8.1: Street signs at the Company camp.
Source: Photograph by Peter D. Dwyer, 2013.

The physical structure of the camp reinforced the sense of attributed rank and differential authority that prevailed there. Thus, the quality of accommodation rapidly reduced from the air-conditioned, generously fitted, large tent where the camp manager worked and slept, through the smaller but private tents for lesser ranked senior staff—geologists, helicopter pilots, medic and others—to the communal, dormitory-like tent where Papua New Guinean 'foremen' were housed. And when the site was fully operational, with seismic or geo-surveys underway and a hundred or more outsiders either in residence or back and forth between field and base camp, senior staff—professionals—were served high quality and varied meals in one well appointed mess while 'foremen' and labourers were fed a less varied diet at a well separated, and much less well appointed, mess that could not be accessed via a covered walkway.

Time too was carefully structured. There was a compulsory early morning meeting to discuss and reinforce issues of safety, and fixed times when meals were served, when people should be seen to be working and when they were free to wash or to rest. The places people occupied, the facilities they were provided with at those places, the paths they followed to move between places and the times at which they should be found at these places were all given in advance. They were preordained. There was little room for contingency and no expectation that individuals—at any level—would not conform.

The camp was both a gendered and hierarchical space. With the exception of a few local women who worked as 'laundry girls' there were no women. At no time, in our experience, was a woman included among the Company workers who came from elsewhere. Hierarchy, as noted, was evident in the physical structure of the space but it was evident, too, in the way people dressed—helmet colour reflected status in field situations—and in what they did, and in who they instructed or were instructed by with respect to those tasks. White staff were ranked most highly, locally recruited staff held the lowest positions. But even the former lacked independence. They were in daily contact by satellite phone, radio or email with senior Company officials who were based elsewhere, at offices in Moro at Lake Kutubu, Yavo on the Strickland River, Port Moresby, Sydney or Calgary. It was unnamed 'bosses' elsewhere who determined who did what and when it should be done, who determined when supplies of food and equipment would arrive, when there would be a change-over in resident camp manager, and when the camp would enter a 'standby period' or be 'demobbed'. The lives of those who worked at the Suabi camp were, to a large extent, controlled by distant outsiders who, in the final analysis, not even the most senior camp employee was likely to have ever met. Indeed, it was rare that any of the expatriate men who worked there met each other, or knew how the others lived, in contexts outside the camp. And, to the extent that the exploration camp was focal for local Kubo and Febi people, providing some immediate employment and implying future possibilities, so too their lives were increasingly beholden to those distant, unknown outsiders.

The camp itself was not unchanging. Gama ran the camp but, at different times, did so for clients—Oil Search or Talisman—who, though material evidence of safety was a major consideration to both, prioritised different aspects of structure and performance. And at times when the camp was on hold, with a small staff and no exploratory work underway, some camp managers were more relaxed than others about enforcing hierarchical differences between workers. One acknowledged to us his reliance on the knowledge and experience of his 'deputy'—a Papua New Guinean man—and suggested that, when no exploration teams were based at the camp, there was no need for highly paid white staff like him. What had to be done could be easily accomplished by men such as the one to whom he himself turned for guidance.

For Company representatives, then, both the social world and the material world were grasped largely in categorical terms. They lived in a highly structured environment, with space and time, roles and expectations clearly defined: a camp where the place and purpose of each tent was known and movement between them followed clearly defined paths; where roles were marked by the colour of a helmet; and where the clock marked what should be done next.

All this, Kubo and Febi people saw, and talked about, every day when exploration camps were based at Suabi. It is what many of them experienced directly, when they themselves were employed either at the camp on in the field. It was part of daily experience through the years 2006 to 2009, when a base camp at Suabi serviced a long phase of work in the Juha area. Their observation of and encounters with the camp, in turn, had multiple effects. They revealed the material disadvantage of local people relative to those who came from outside and demonstrated that what was once rumoured was, indeed, the state of the world beyond the limits of their land and their previous imaginings. They revealed ways in which the lives—indeed the social being—of outsiders were structured and performed: the ways in which gender and status were explicit organisational props; workers were more valued for their performance of designated tasks than for their capacities as persons; the physical structure of the camp came to define the site and its parts without reference to a broader context; and employees at all levels were both subject to the dictates of seniors and expendable and replaceable if they failed to conform. Those observations and encounters revealed, too, the limited place of relational understandings and actions in the lives of those who came to their land in search of oil or gas. And they revealed that those people, though seemingly powerful, were themselves subject to the authority of unseen others.

While people at Suabi may have strived to know Company representatives as persons—to know their personal names, and engage with them as particular 'best friends'—they could not avoid the recognition that those men, as camp manager or medic or mechanic, also 'personated' Company and the bureaucratic leviathan it was (compare Golub 2014). Everyday encounters in and around the camp revealed directly, through grounded experience, that the sociality of Company—'Company' as an entity—

was underwritten by reification, categorisation, commensurability and anonymisation; that the 'things' of the world of Company and the ends to which those 'things' were put were at a far remove from the ontological and epistemological persuasions that had once underpinned the experience of local people. It made explicit the more tentative experiences and understandings that were emerging in contexts of locally-based missions and school. It revealed that categorical expressions of ontology and epistemology could be extrapolated to all domains of living. And it revealed, too, that desires might best be satisfied, or at least made tangible, by conforming to those categorical imperatives—by performing as the outsiders they hosted performed.

Political Ontology and Identity

Ontologies change. When peoples with different ontologies meet each may influence the other though, not unusually, both the relationship and the exchange are asymmetrical. Diane Austin-Broos (1996) wrote of two 'ways of being Aranda'—of living simultaneously, as the people themselves said, by 'Aranda law and God's law'—at Hermannsburg in Australia. 'Aborigines', she wrote, 'change and reposition practice in response to European incursions' (ibid.: 5). They do so 'both as a conscious practice and as an unconscious one … as [n]ew elements of practice and knowledge are assumed [and] positioned and repositioned within an Aboriginal circumstance' (ibid.). And, always, identity is at stake. '"Two-laws" talk refers to ontologies, but also marks a passage between them. The identities this talk articulates are ethnic identities, modes of historical consciousness based on a social experience of difference that is also hierarchical' (ibid.: 6). 'Ontology', she wrote, 'is not something beyond practice … but rather a specifying integration of practice that defines a world that can also be changed or radically undermined' (ibid.: 18).

There are places and times, however, when the encounter between parties whose originating ontological persuasions differ erupts as dispute. Where tensions arise, issues of identity may be foregrounded and, even though 'new elements of practice and knowledge' have been incorporated into ways of being by one or other of the parties, what has seemingly changed is put aside in favour of what has seemingly persisted (Keesing 1989). Disputation promotes discourses of difference and these, as Blaser (2009b: 11) wrote, often may be underwritten by 'the entities that make up a particular world or ontology'.

In the late 1990s, Yshiro people of Northern Paraguay came into conflict with inspectors of the National Parks Direction over an issue of sustainable, though commercial, hunting. The former asserted that sustainability necessarily entailed maintaining far-reaching reciprocal relations with others, while the latter asserted that both 'Yshiro and nonindigenous hunters were actively disregarding the agreed-on regulations, thereby turning the program into "depredation" and "devastation"' (ibid.: 10). To Yshiro the 'human-to-human interface' was central to their understanding of conservation; to scientifically-minded bureaucrats it was 'human-to-animal relations' that were central (ibid.: 14). The Yshiro prioritised qualitative considerations; the outsiders prioritised quantitative considerations. As Blaser (ibid.: 11) wrote, an apparent agreement to adopt 'sustainable' practices negated realisation 'that "animals"—and, by extension, the world(s) they are part of—were radically different entities for the Yshiro and for bureaucrats and experts involved in the hunting program'. To the former, animals are agentive beings entangled in relations of reciprocity with humans and cohabiting a world or cosmos—the *yrmo*—that is, itself, 'governed by ... the mutual dependence of all that exists' (ibid.: 13). To the latter, animals lack 'volition of their own' but, rather, are 'conceived as objects' standing apart from, though ultimately measurable and manageable by, responsible humans. It was this ontological difference, Blaser argued, that was at the heart of the dispute and awkward negotiations he described.[9] This difference, however, came to the fore at a time when the encounter of Yshiro people with government, Christian missions, loggers and military had persisted for more than a century; 'in the early decades of the 20th century ... [Yshiro] were more or less forcefully incorporated as a cheap labor force into logging camps', and later, when the logging companies collapsed, were resettled at missions and cattle ranches as their own lands were sold to speculators (ibid.: 12). They were not uncontaminated by

---

9   Andreas Roepstorff (2003) has discussed the incommensurability of understandings of 'overfishing' by commercial fishermen of Greenland and the Danish fisheries scientists who advise on management. To the former, 'overfishing' connotes the behaviour of those who take more than they need; to the latter, 'overfishing' connotes an overall level of take, irrespective of the behaviour of particular fishermen, that jeopardises maintenance of viable stock. He observes, as well, that where the fishermen understand fish as like 'non-human persons' that may be known only by interaction and engagement, to the fisheries scientists fish are 'objects' amenable to abstract quantitative analysis (ibid.: 131–3). These divergent understandings, for both parties, entangle objectivity and subjectivity and arise in practice. To the fishermen and the fisheries scientists 'fish' and 'overfishing' might be understood, in Bruno Latour's (1999) terminology, neither as 'facts' nor 'fetishes', but as 'factishes'. The same may be said of the notions of 'animal' and 'sustainability' to Yshiro and the bureaucrats and experts with whom they come into conflict.

those experiences. They had put new social structures in place—a Yshiro federation, for example—and learned new ways of resolving new problems as evidenced, for example, by their success in obtaining support from a European Union sustainable development fund (ibid.: 10).

Alex Golub's (2006, 2014) detailed exploration of the encounter—often fraught, always ambiguous—between Ipili-speaking Porgeran people of highland PNG and the enormous gold mine established on their land in the late 1980s also may be appreciated in a frame of political ontology. By 1992, its second year of production, Porgera was the third-largest gold mine in the world and, by 2000, had spent more than PGK13 million on salaries and wages—much directed to Porgeran people—and PGK20 million as donations to local groups. In addition, the Porgerans have received tens of millions of kina as compensation payments for damage to their land and as royalties from the sale of gold and, through the government, have been provided with roads, a hospital, support for school fees and other services.

The Porgerans, however, were by no means satisfied. 'Ten years into what was supposed to be an age of effortless wealth and health, they felt cheated by what the mine had wrought and deeply entitled to more than they had received' (Golub 2014: 30). They were, by this time, experienced and forceful negotiators with access to, and the funds to pay for, legal advice. They knew that documentation—both their own and that provided by the mine—could both legitimise demands or promises and, at times, be constructed to reveal or conceal intended actions. They knew that delaying tactics could sometimes jeopardise the financial position of the mine and encourage concessions in their favour. They had, themselves, assumed many of the organisational trappings of bureaucratic and corporate bodies. But they were not united. There were factions within Porgeran society with competing agendas and, thus, employing different 'facts' and different strategies in negotiating with each other and with representatives of the mine.

Golub (2014: 12–7) depicts both the mine and the Porgerans as leviathans, but leviathans of different sorts. The former is 'leviathan-as-bureaucracy' with 'professional, dedicated and disciplined people acting in accordance with predetermined rules and regulations'. It is 'regimented, efficacious, [and] expansive in its power'. The latter is leviathan-as-cosmology where 'standardization, legibility, and simplification'—in the Porgeran case of, for example, '"customary law," "kinship," and "clan systems"'—underwrite a sense of order. And it is in this difference, at base, that there always has been an unresolved disjunction between the understandings, and the concomitant intentions and strategies, of the mine, on the one hand, and Porgerans on the other. Neither party fully comprehends the other. Representatives of the mine understand themselves to be 'doing good' for the company, for PNG and for local people but do not recognise, or acknowledge, the intricacies and never-ending flux of local perceptions of kinship and land ownership and the ways in which these are deployed in practice or debate (compare Filer 1997). They expect both compliance and gratitude, but do not get them. They incline to the opinion that local people are greedy, ignorant and unwilling to appreciate the many benefits of health, education, housing, etc. that have flowed from the presence of the mine. Porgerans understand themselves to be presenting objective accounts of kinship and landownership but do not recognise, or acknowledge, that these accounts remain deeply committed to a dynamic, and always contingent, relational ethos. They expect both comprehension and generosity, but do not get them. They incline to the opinion that officers of the mine, whatever their status, are greedy, deceitful and unwilling to act according to the rights, and ongoing disadvantage, of landowners. The mine and the Porgerans have been engaged in a political struggle—one which, Golub (2014: 208–13) concluded, by the mid-2000s the Porgerans had lost—that resisted resolution because the two central, though ultimately indefinable, participants knew the world in different ways.

On multiple counts, the scale of the encounters between local people and resource extracting companies differs greatly between the operations focused at, respectively, Porgera and Juha. The Porgerans were contacted, brought under colonial administration control, and initiated access to and interpretations of Western ways, in the late 1930s and 1940s, 30 years earlier than Kubo and Febi. In 2000, the Porgerans probably numbered about 23,000 people with, by PNG standards, a moderately high population density of around 36 people per km^2 (Jacka 2015: 39, 46, 181–2); by 2010, Kubo and Febi numbered about 1,500 people with an overall population density of less than one person per km^2.[10] The Porgerans

---

10 In both the Kubo-Febi and Porgeran cases much of the area of land over which they hold rights is not used or very seldom used. At Porgera, in-migration prompted by mining has seen a great increase in population. Most of the immigrants are not Ipili-speakers. Between 1990 and 2004 the number of people living within the area of the Special Mining Lease increased from about 2000 to about 15,000 (Banks 2008: 30; see also Jacka 2015).

formalised agreements with company and government in the 1980s; agreements with Kubo and Febi had not been finalised by mid-2014. Porgerans have received a hundred million or more kina in royalties and grants since mining commenced in the early 1990s. Kubo and Febi are yet to receive royalty payments from gas wells on the land of the latter people and have received little more than two million kina as Infrastructure or Business Development Grants. Porgerans own the high rise building in Port Moresby where Placer Niugini—the operator of the mine—located its headquarters (Golub 2006: 268); the only property owned by Kubo or Febi people beyond the limits of their own lands is a house in Port Moresby. Additionally, while the visible imprint of quarrying on the land at the Porgera mine is vast and extensive, the visible imprint of the wells sunk at Juha is minor, highly localised and will remain small. And, tellingly, while the situation at Porgera has been often one of contention and conflict, at Suabi there has been neither overt contention nor conflict in the experience of local people and the companies they have hosted. For the Porgerans, issues of identity have come to the fore. At Suabi, to the middle of 2014, though people desired more than they had, felt disadvantaged and strove to improve their lot, they did not respond as though their sense of identity was being challenged.

Continuity and Change

Through the past 30 years Kubo and Febi people have changed a great deal. Population has doubled, lifespan has been extended, formerly dispersed longhouse communities have come together, access to money and formal education has increased, many people are now relatively fluent in Tok Pisin and English where previously most had competence in only their natal language and, perhaps, the language of their immediate neighbours, and travel beyond their own territories is now a common experience. All these changes were facilitated by the arrival of missionaries and of health and education services, and by the intermittent visits of outsiders in search of minerals, oil or gas. But these easily detected changes are surface appearances only. Of greater significance were the gradual and continuing shifts in the ontological and epistemological foundations of people's lives. In the context of new experiences, and in pursuit of the opportunities these experiences promised, people acted on the world in different ways. And, in so doing, they came to know things and to employ those things in different ways. In many domains of their lives, but not in all and not in all contexts, they placed less emphasis upon the ways in which they and the 'things' of their world—other people, other beings and material items—were entangled one with another in far-reaching, never dissolvable, networks of relations, and came to place more emphasis on those things as 'objects' that stood apart from other objects, categorisable by attributes possessed in common with other objects of the same kind. Gradually, a categorical imperative infiltrated the material and temporal expressions of their lives (house and village structures, scheduling of activities), exchanges of women and pigs, the ways they named people, and the ways they conceptualised and, most recently, recorded the form and organisation of social groups and the connections these groups had with land. And in parallel with these changes—indeed as party to them—people themselves were transformed as, increasingly, they chose to act as individuals, beholden to personal interests and desires and less obliged to always concede to the relational ethos within which they continued to be embedded.

But all these changes, whether surface or deep, were not forced upon Kubo and Febi. They learned that life was different elsewhere. Outsiders came to their land and, in their own ways of living, both revealed something of those differences and enhanced desire for what might be offered by embracing such differences. The changes that then followed were, in large part, an outcome of the actions of the people themselves. There was no external force that drove those changes. In the ways they talked and acted, missionaries, company employees, government officers, schoolteachers, health workers and anthropologists presented models of what could be possible, and thereby enhanced desire and expectations. But, sometimes despite their own intentions, those outsiders were not the ultimate cause of changes. The changes we have described arose internally, as expressions of the agency of the people themselves; as outcomes of the conscious choices of people 'trying to solve their problems and achieve their aims' (Strathern and Stewart 2004: 160).

There was, however, much about the way in which Kubo and Febi people interacted with one another, and with the environment that was the source of their livelihood, that persisted through this time. With respect to garnering needed resources from the environment, Kubo had been risk averse; that is, they favoured security of supply over security of tenure and, to this end, tracked shifting patterns of resource availability by moving between different modes of acquiring that which was currently available. Immediate response to changing circumstances was both usual and expected; the need to redirect primary modes of appropriation was not a cause for alarm. When outsiders came and provided access to money and the things and opportunities that this made possible then, for a time, people redirected attention to these new possibilities. But when the outsiders departed, as they usually did, people returned to conventional ways of living; they gardened more, hunted more and processed more sago. They did not, however, lose their desire for that which they had recently experienced.

Again, as discussed in Chapter 7, Kubo people acted on the environment as a world of affordances that 'gave' without a requirement that they, in turn, reciprocate. They understood, however, that risk was entailed, that there were forces at play beyond their control. The environment was unpredictable—it could give what was not wanted or, at times, withhold what was wanted. Reciprocity was not necessary; respect and caution were. When outsiders came, bringing the possibility that emerging desires might be satisfied, local people sought to engage with them as persons. But, in as much as those newcomers failed to respond as expected of persons, they revealed themselves to be 'other'. Increasingly, they were positioned within the frame of an environment that 'gave' and eventually, as with petroleum companies, of an environment from which people could 'take'.

And, finally, through all this time, there remained a deep concern with sorcery, with the possibility of random assault from *hugai* sorcerers that lurked in the forest or came from elsewhere, of unwittingly giving offence to a person you knew and being targeted by them by means of *bogei* (parcel) sorcery, or of being entrapped by new, though little-understood, forms of sorcery that were encountered beyond home territory or arrived with outsiders. People at Suabi were always alert to the need to do what they could to offset the likelihood that someone would make them the target of sorcery or conclude that they were practising sorcery. As we wrote of our early experiences at Gwaimasi: 'Among Kubo a concern with sorcery was an abiding reality and those with whom one had day to day dealings were potentially most threatening', and though 'departure could defer both the threat and the accusation of sorcery ... it was through food-sharing that everyone in a community sought to reduce the likelihood that they themselves were in danger' (Dwyer and Minnegal 1992a: 48). With the proviso that departure is no longer an easy option—people are now less dispersed across the landscape, with most living in close association at the large village of Suabi—these concerns and responses persist to the present time. It is in the public space of feasts that people declare by their behaviour that they feel no antagonism towards others, and thus are neither inclined to sorcery nor deserving of ensorcelment. In the ways people share food at these events they emphasise equivalence, personal neutrality and the relational bonds that—in the past, in the present and into the future—have and will sustain them.

All that had changed, all that was changing and all that had persisted provided the frame within which the people who lived at Suabi in the years 2011 to 2014 sought ways to establish their credentials as legitimate beneficiaries to the wealth they expected to come from Juha. The activities of the PNG LNG Project on and near their land held out this promise of untold wealth and simultaneously, through the practices of its representatives, seemed to reveal what should be done to bring that promise to fruition. This was the context, then, in which Kubo and Febi people came to put in place new conceptualisations of the world of 'things', with respect to both what they were and how they might be employed. It was in this context that they drew up lists of members of ILGs, or devised logos to stand for different configurations of people and, thereby, reshape social form. It was in this context—the 'shadow of the liberal diaspora' (Povinelli 2001: 320)—that they sought to be radically different, to render themselves visible to the state and to be heard. And it was in this context that they were deeply engrossed in navigating a future that, because their experience of the present was ever-changing and thus lacked clear guide posts to the way ahead, was itself uncertain.

The task was forbidding for, despite all that changed, to Kubo and Febi people, on the one hand, and Company, on the other, there remained an insurmountable difference in the worlds of 'things' to which each was committed. In this domain there was no 'social commensuration' (Povinelli 2001). For Company, the abstractions of science and the global market prevailed. For Kubo and Febi, despite the increasing place of categorical imperatives in their lives, the grounded experience of exchange relations and the sense of environment as giving persisted. For Company, the trappings of bureaucracy prevailed; for Kubo and Febi, the cosmological could not be put aside. To each party, different 'things' were at stake. But the outcome was that, in their encounter, and in the long term, Kubo and Febi were obliged—unwittingly perhaps—to concede more than Company. They were positioned by the representatives of an ontology that was grounded in certainty and who understood other life worlds to be either inferior variants of their own or striving to participate in, and join, their own.

Indeed, for people at Suabi, time itself had to be reimagined. In our earliest years at Gwaimasi, when we travelled with people, we would often ask 'whose sago palm is this?' or 'whose grove of fruit pandanus is this?' or 'whose land is this?' Sometimes, especially if the person we asked was unmarried or as yet without children, they might declare the palm, or the fruit pandanus or the land to be their own (Minnegal and Dwyer 1999: 66). What they did not do is attribute the item in question to an ancestor, to their father or to another person's father or grandfather. But most often, they named a child, sometimes a boy, sometimes a girl, as the person associated with the plants or the land to which we had directed attention. They did so not as an assertion of personal ownership—the land had not been apportioned among the children of a family—but rather as a way of inscribing the named child onto the land. In these contexts, and again when they renamed parents after a child born to those parents, their orientation was toward the future. With respect to both land and interpersonal relationships the meaning of the present was sought by reference to contingencies of the future. Identity was not understood as constrained by past actions but, rather, by what one might do next. Yet, while 'what is done next' might reconfigure the relational field and thus change the identity of the actor, it was expected that the relational embeddedness of current practice would be reproduced. But this is changing. Past, present and future are being reconceptualised as an inevitable, and abstract, sequence; there is increasing adherence to calendrical time—fixed market days, Easter, Christmas, birthdays—in patterning activities. Rights to land and resources are now asserted by declaring ancestral connections, and the benefits that may flow from them are expected to be delivered in the near future to those who are established as rightful owners, irrespective of what they themselves have done. It is neither expected, nor desired, that the future will reproduce the present, let alone the past. The future, however, now imagined as a very different—and undoubtedly better—place, is problematic. It is a place that beckons, a place of as yet unfulfilled promises, a place that may be reached only by navigating the ever-shifting landscape that lies between present circumstance and present desire.

In all these ways, therefore, a new world is emerging for Kubo and Febi people and they themselves are emerging as new kinds of subjects within that world.[11]

Coda: Then and Now

Infants are everywhere at Suabi. There are so many of them. But the explosion in numbers has not detracted from the overwhelming attention they receive from their parents, their older siblings and from anyone else who is passing by. A baby will be lifted from its mother's arms, carried away, shown to others, talked to and caressed. A mother, father or sibling will point to a garden plant, a pig, a chicken, an aeroplane overhead, holding the child towards the object of attention and saying, as appropriate, 'Look, banana, look', 'Look, your pig, look'. Even before the child can hold its own head steady, they encourage it to look. And, at first, they fail to get the response they seek. This takes time. As one father told us when his four-month-old son failed to focus on a passing chicken: 'He is seeing, but he is not yet looking'. The child was not yet attending to what was going on in the world around him.

In May 1987 we joined a group from Gwaimasi who were visiting Gugwuasu, six hours travel away by canoe, raft and foot, to celebrate completion of a new longhouse and village site. It was a relaxed day. Some of us walked down-river to a place where we were ferried by canoe to the small hamlet at Sosoibi on the opposite bank. Some travelled by canoe and raft, transporting more than 100 kg of dried meat—pig, cassowary, cuscus, bandicoot, fish, sago grubs—that had been prepared through the preceding few weeks. We met with other people at Sosoibi, gossiped and shared food. There was no hurry; we would sleep here tonight and continue to Gugwuasu in the morning. A few easy hours walking through the forest would get us there, and we did not want to arrive until late afternoon. While we rested at Sosoibi, Digati made a rattle. She had saved the claws of a species of small crayfish that is favoured as fish bait. She threaded these together as a bundle in such a way that, if jostled, they

---

11   In their recent account of the lives of some Chambri people, Deborah Gewertz and Frederick Errington (2016: 349–50) show how encounters with 'capitalism have demanded serious readjustments in the nature of Chambri habitus'. Reflecting on observations of Marshall Sahlins concerning the Eskimo and Hawaiians they write that, despite decades of change, 'the Chambri are still there and are still Chambri; yet they are definitely and fundamentally not their (grand)father's Chambri'. They are (ontologically) new subjects.

would click one against another. When the rattle was finished, Digati attached it to the string bag in which she carried her five-month-old infant son Gai. It was a miniature version of the rattle worn by Kubo men when, fully painted and costumed, they danced through the night to celebrate the move to a new longhouse or at curing ceremonies (Dwyer and Minnegal 1988). The rattle is heard by the spirits of the dead who come to give support to the living. This is what Digati intended for Gai. She was directing the attention of the spirits to her infant child. She was calling on their beneficence in shaping the future of her son.

In 2014, at Suabi, we saw a lot of Alice and her infant daughter Kamari. Alice was having an unhappy time living in the house with her husband's parents. She often needed to escape. And our open living space caught the breeze and was cooler than the enclosed house in which she lived. While we wrote notes or typed lists, she would suspend Kamari's string bag from a beam and wander away to complete other tasks. Above Kamari, hanging from a length of home-spun string and swinging gently, was a one kina coin; the Papua New Guinean coin that has a hole through the centre. It was close enough to be seen by Kamari but she was not yet old enough to reach out and touch it (Fig. 8.2).

At Sosoibi, Digati was calling on the spirits of the dead to care for her son in the present and into the future. She was invoking the past. Twenty-eight years later, at Suabi, another young mother sought a profoundly different way in which the future of her infant daughter might be secured. She did not seek to draw the attention of beneficent spirits to her child. She did not call on the past. She directed Kamari's attention to an iconic representation of that desired future. In doing so she sought to clear the path that her child would need to navigate.

Figure 8.2: 'Directing attention' and navigating the future.
Source: Photograph by Monica Minnegal, 2014.

References

Adhikari, A., A. Sen, R.C. Brumbaugh and J. Schwartz, 2011. 'Altered Growth Patterns of a Mountain Ok Population of Papua New Guinea Over 25 Years of Change.' *American Journal of Human Biology* 23: 325–332 (doi.org/10.1002/ajhb.21134).

Allen, B.J., 1990. 'George Arthur Vickers Stanley (1904–1965).' In *Australian Dictionary of Biography* Vol. 12. Carlton: Melbourne University Press. Viewed 24 April 2016 at: adb.anu.edu.au/biography/stanley-george-arthur-vickers-8624

Anderson, B.R.O'G., 1991. *Imagined Communities: Reflections on the Origin and Spread of Nationalism*. New York: Verso.

Anderson, J.L., 1970. *Cannibal: A Photographic Audacity*. Sydney: A.H. & A.W. Reed.

Anon., 1887. 'Recent Explorations in New Guinea.' In *Report of the Fifty-Sixth Meeting of the British Association for the Advancement of Science held at Birmingham in September 1886*. London: John Murray.

——, 2009. 'Juha Landowners Sign LNG Deal.' *Post-Courier*, 9 December.

APC (Australasian Petroleum Company), 1940. *Oil Exploration in Papua and the Mandated Territory of New Guinea*. Melbourne: National Press.

Appadurai, A., 1986. *The Social Life of Things: Commodities in a Cultural Perspective*. Cambridge: Cambridge University Press (doi.org/10.1017/CBO9780511819582).

——, 1990. 'Disjuncture and Difference in the Global Cultural Economy.' *Public Culture* 2: 1–24 (doi.org/10.1215/08992363-2-2-1).

Austin-Broos, D.J., 1996. '"Two Laws", Ontologies, Histories: Ways of Being Aranda Today.' *The Australian Journal of Anthropology* 7: 1–20 (doi.org/10.1111/j.1835-9310.1996.tb00334.x).

Bacalzo, D., B. Beer and T. Schwoerer, 2014. 'Mining Narratives, the Revival of "Clans" and other Changes in Wampar Social Imaginaries: A Case Study from Papua New Guinea.' *Journal de la Société des Océanistes* 138–139: 63–76 (doi.org/10.4000/jso.7128).

Bainton, N.A., 2010. *The Lihir Destiny: Cultural Responses to Mining in Melanesia.* Canberra: ANU E Press.

Bainton, N.A., C. Ballard and K. Gillespie, 2012. 'The End of the Beginning? Mining, Sacred Geographies, Memory and Performance in Lihir.' *The Australian Journal of Anthropology* 23: 22–49.

Banerjee, S.B., 2000. 'Whose Land is it Anyway? National Interest, Indigenous Stakeholders, and Colonial Discourses: The Case of the Jabiluka Uranium Mine.' *Organization Environment* 13: 3–38 (doi.org/10.1177/1086026600131001).

Banks, G., 2008. 'Understanding "Resource" Conflicts in Papua New Guinea.' *Asia Pacific Viewpoint* 49: 23–34 (doi.org/10.1111/j.1467-8373.2008.00358.x).

Barclay, R.I., 1971a. 'Nomad Patrol Report 16 of 1970/71: Western District.' Territory of Papua and New Guinea.

——, 1971b. 'Nomad Patrol Report 4 of 1971/72: Western District.' Territory of Papua and New Guinea.

——, 1972. 'Nomad Patrol Report 3 of 1972/73: Western District.' Territory of Papua and New Guinea.

——, 2012. 'In Pursuit of Cannibals.' *Quadrant Magazine* 56: 92–98.

Barth, F., 1971. 'Tribes and Intertribal Relations in the Fly Headwaters.' *Oceania* 41: 171–191 (doi.org/10.1002/j.1834-4461.1971.tb01150.x).

——, 1987. *Cosmologies in the Making: A Generative Approach to Cultural Variation in Inner New Guinea*. Cambridge: Cambridge University Press (doi.org/10.1017/CBO9780511607707).

Bashir, M., 2010a. 'Senior Officers Face Fraud.' *Post-Courier*, 28 May.

——, 2010b. 'Sinalis "Lost Tribe of LNG Project".' *Post-Courier*, 14 October.

——, 2010c. 'Donigi Book Highlights Clan's Bid in LNG Project.' *Post-Courier*, 2 November.

Bateson, G., 1979. *Mind and Nature: A Necessary Unity*. New York: Dutton.

Bauerlin, W., 1886. *The Voyage of the Bonito: An Account of the Fly River Expedition to New Guinea*. London: Gibbs, Shallard.

Beek, A.G. van, 1987. The Way of all Flesh: Hunting and Ideology of the Bedamuni of the Great Papuan Plateau (Papua New Guinea). Leiden: University of Leiden (PhD thesis).

Bell, J.A., 2009. 'Documenting Discontent: Struggles for Recognition in the Purari Delta of Papua New Guinea.' *The Australian Journal of Anthropology* 20: 28–47 (doi.org/10.1111/j.1757-6547.2009.00002.x).

——, 2016. 'Dystopian Realities and Archival Dreams in the Purari Delta of Papua New Guinea.' *Social Anthropology/Anthropologie Sociale* 24: 20–35 (doi.org/10.1111/1469-8676.12285).

Besasparis, B.A., 1959. 'Kiunga Patrol Report No. 9 of 59/60: Western District.' Territory of Papua and New Guinea.

Bestor, T.C., 2001. 'Supply-Side Sushi: Commodity, Market, and the Global City.' *American Anthropologist* 103: 76–95 (doi.org/10.1525/aa.2001.103.1.76).

Biersack, A., 1999. 'The Mount Kare Python and his Gold: Totemism and Ecology in the Papua New Guinea Highlands.' *American Anthropologist* 101: 68–87 (doi.org/10.1525/aa.1999.101.1.68).

Bird-David, N., 1990. 'The Giving Environment: Another Perspective on the Economic System of Gatherer-hunters.' *Current Anthropology* 31: 189–196 (doi.org/10.1086/203825).

——, 1994. 'Sociality and Immediacy: Or, Past and Present Conversations on Bands.' *Man* (N.S.) 29: 583–603 (doi.org/10.2307/2804344).

——, 1999. '"Animism" Revisited: Personhood, Environment, and Relational Epistemology.' *Current Anthropology* 40: S67–S91 (doi.org/10.1086/200061).

——, 2004. 'No Past, No Present: A Critical-Nayaka Perspective on Cultural Remembering.' *American Ethnologist* 31: 406–421 (doi.org/10.1525/ae.2004.31.3.406).

——, 2006. 'Animistic Epistemology: Why Do Some Hunter-Gatherers Not Depict Animals?' *Ethnos: Journal of Anthropology* 71: 33–50 (doi.org/10.1080/00141840600603152).

Blaser, M., 2009a. 'Political Ontology.' *Cultural Studies* 23: 873–896 (doi.org/10.1080/09502380903208023).

——, 2009b. 'The Threat of the Yrmo: The Political Ontology of a Sustainable Hunting Program.' *American Anthropologist* 111: 10–20 (doi.org/10.1111/j.1548-1433.2009.01073.x).

——, 2013. 'Ontological Conflicts and the Stories of Peoples in Spite of Europe. Toward a Conversation on Political Ontology.' *Current Anthropology* 54: 547–568 (doi.org/10.1086/672270).

Bourdieu, P., 2001. 'Television.' *European Review* 9: 245–256 (doi.org/10.1017/S1062798701000230).

Bowman, P., 2004. 'Nomad Report: Australian Doctors International, 24 December 2004.' Viewed 9 September 2014 at: www.yumpu.com/en/document/view/8038453/nomad-report

Brown, M.F., 1996. 'Forum: On Resisting Resistance.' *American Anthropologist* 98: 729–735 (doi.org/10.1525/aa.1996.98.4.02a00030).

Bryan, J.E. and P.L. Shearman (eds), 2015. *The State of the Forests of Papua New Guinea 2014: Measuring Change over the Period 2002–2014.* Port Moresby: University of Papua New Guinea.

Busby, C., 1997. 'Permeable and Partible Persons: A Comparative Analysis of Gender and Body in South India and Melanesia.' *The Journal of the Royal Anthropological Institute* (N.S.) 3: 261–278 (doi.org/10.2307/3035019).

Business Advantage, 2014. 'Papua New Guinea's PNG LNG Project: Who Gets the Money?' 27 May 2014. Viewed 31 May 2014 at: www. businessadvantagepng.com/papua-new-guineas-png-lng-project-gets-money

Busse, M. and V. Strang, 2011. 'Introduction: Ownership and Appropriation.' In M. Busse and V. Strang (eds), *Ownership and Appropriation*. Oxford: Berg (ASA Monograph 47) (doi.org/10. 1484/m.mcs-eb.4.3002).

Calder, D.G., 1953. 'Lake Murray Patrol Report No. 1 of 1953/54: Western District.' Territory of Papua and New Guinea.

Carrithers, M., M. Candea, K. Sykes, M. Holbraad and S. Venkatesan, 2010. 'Ontology Is Just Another Word for Culture: Motion Tabled at the 2008 Meeting of the Group for Debates in Anthropological Theory, University of Manchester.' *Critique of Anthropology* 30: 152–200 (doi.org/10.1177/0308275X09364070).

Cawthorn, W.A., 1970. 'Nomad Patrol Report 13/69–70: Western District.' Territory of Papua and New Guinea.

Chambers, M., 2016. 'Huge PNG Gas Find Could Power LNG Plant.' *Australian Business Review*, 30 December.

Chandler, J., 2011a. 'Little Common Ground as Land Grab Splits a People.' *Sydney Morning Herald*, 15 October.

———, 2011b. 'PNG's Great Land Grab Sparks Fightback by Traditional Owners.' *The Age*, 14 October.

Clancy, D.J., 1948. 'Daru Patrol Report No. 2 of 1947/48: Western District.' Territory of Papua and New Guinea.

Coffey Natural Systems, 2009. *PNG LNG Project: Environmental Impact Statement* (9 volumes). Report to ExxonMobil and joint venture partners.

Comaroff, J. and J. Comaroff, 2009. *Ethnicity, Inc.* Chicago: University of Chicago Press (doi.org/10.7208/chicago/9780226114736.001.0001).

Compagno, L.J.V., W.T. White and P.R. Last, 2008. '*Glyphis Garricki* Sp. Nov., a New Species of River Shark (Carcharhiniformes: Carcharhinidae) from Northern Australia and Papua New Guinea, with a Redescription of *Glyphis glyphis* (Müller & Henle, 1839).' In P.R. Last, W.T. White and J.J. Pogonoski (eds), *Descriptions of New Australian Chondrichthyans*. Hobart: Commonwealth Scientific and Industrial Research Organisation (Marine and Atmospheric Research Paper 022: 203–225).

Coumans, C., 2011. 'Occupying Spaces Created by Conflict: Anthropologists, Development NGOs, Responsible Investment, and Mining.' *Current Anthropology* 52: S29-S43 (doi.org/10.1086/656473).

Coxworth, B., 2013. 'Review: Waka Waka Power Solar Lamp and Device Charger.' Viewed 22 February 2016 at: www.gizmag.com/review-waka-waka-power-solar-lamp-phone-charger/27543

Craig, B., 2013. 'How Karius Found a River to the North: The First 1927 Attempt to Cross New Guinea from the Fly to the Sepik.' Viewed 31 May 2017 at: www.uscngp.com/papers/

Craig, M.S. and K. Warvakai, 2009. 'Structure of an Active Foreland Fold and Thrust Belt, Papua New Guinea.' *Australian Journal of Earth Sciences* 56: 719–738 (doi.org/10.1080/08120090903005360).

Curry, G.N. and G. Koczberski, 2009. 'Finding Common Ground: Relational Concepts of Land Tenure and Economy in the Oil Palm Frontier of Papua New Guinea.' *The Geographical Journal* 175: 98–111 (doi.org/10.1111/j.1475-4959.2008.00319.x).

Cuthbert, N., 2009. 'The Hela Prophesy.' Viewed 28 October 2015 at: malumnalu.blogspot.com.au/2009/11/hela-prophesy.html

Demian, M., 2006. '"Emptiness" and Complementarity in Suau Reproductive Strategies.' In S.J. Ulijaszek (ed.), *Population, Reproduction and Fertility in Melanesia*. Oxford: Berghahn Books.

Denham, T., A. Bedingfield and U. Gilad, 2009. '4.2.1 Juha to Hides.' In L. Goldman (ed,), *Papua New Guinea Liquefied Natural Gas Project: Social Impact Assessment 2008*. Report to ExxonMobil Corporation.

Descola, P., 2006. 'Beyond Nature and Culture.' *Proceedings of the British Academy* 139: 137–155.

——, 2013. *Beyond Nature and Culture* (transl. J. Lloyd). Chicago: University of Chicago Press.

Dewey, J. and A.F. Bentley, 1975 (1949). *Knowing and the Known*. Westport (CT): Greenwood Press.

Dwyer, P.D., 1993. 'The Production and Disposal of Pigs by Kubo People of Papua New Guinea.' *Memoirs of the Queensland Museum* 33: 123–142.

——, 1996. 'The Invention of Nature.' In R.F. Ellen and K. Fukui (eds), *Redefining Nature: Ecology, Culture and Domestication*. Oxford: Berg.

——, 2005. 'Ethnoclassification, Ethnoecology and the Imagination.' *Journal de la Société des Océanistes* 120–121: 11–25 (doi.org/10.4000/jso.321).

Dwyer, P.D. and M. Minnegal, 1988. 'Supplication of the Crocodile: A Curing Ritual from Papua New Guinea.' *Australian Natural History* 22: 490–494.

——, 1990. 'Yams and Megapode Mounds in Lowland Rain Forest of Papua New Guinea. *Human Ecology* 18: 177–185 (doi.org/10.1007/BF00889181).

——, 1991. 'Hunting in Lowland Tropical Rainforest: Towards a Model of Non-agricultural Subsistence.' *Human Ecology* 19: 187–212 (doi.org/10.1007/BF00888745).

——, 1992a. 'Ecology and Community Dynamics of Kubo People in the Tropical Lowlands of Papua New Guinea.' *Human Ecology* 20: 21–55 (doi.org/10.1007/BF00889695).

——, 1992b. 'Cassowaries, Chickens and Change: Animal Domestication by Kubo of Papua New Guinea.' *Journal of the Polynesian Society* 101: 373–385.

——, 1993. 'Banana Production by Kubo People of the Interior Lowlands of Papua New Guinea.' *The Papua New Guinea Journal of Agriculture, Forestry and Fisheries* 36: 1–21.

——, 1994. 'Sago Palms and Variable Garden Yields: A Case Study from Papua New Guinea.' *Man and Culture in Oceania* 10: 81–102.

——, 1995. 'Ownership, Individual Effort and the Organization of Labour among Kubo Sago Producers of Papua New Guinea.' *Anthropological Science* 103: 91–104 (doi.org/10.1537/ase.103.91).

——, 1997. 'Sago Games: Cooperation and Change among Sago Producers of Papua New Guinea.' *Evolution and Human Behavior* 18: 89–108 (doi.org/10.1016/S1090-5138(97)00005-6).

——, 1998. 'Waiting for Company: Ethos and Environment among Kubo of Papua New Guinea.' *Journal of the Royal Anthropological Institute* 4: 23–42 (doi.org/10.2307/3034426).

——, 2005. 'Person, Place or Pig: Animal Attachments and Human Transactions in New Guinea.' In J. Knight (ed.), *Animals in Person: Cultural Perspectives on Human-Animal Relations*. Oxford: Berg.

——, 2007. 'Social Change and Agency among Kubo of Papua New Guinea.' *Journal of the Royal Anthropological Institute* (N.S.) 13: 545–562 (doi.org/10.1111/j.1467-9655.2007.00442.x).

——, 2014. 'Where all the Rivers Flow West: Maps, Abstraction and Change in the Papua New Guinea Lowlands.' *The Australian Journal of Anthropology* 25: 37–53 (doi.org/10.1111/taja.12071).

——, 2016. 'Wild Dogs and Village Dogs in New Guinea: Were They Different?' *Australian Mammalogy* 38: 1–11 (doi.org/10.1071/AM15011).

Dwyer, P.D., M. Minnegal and C. Warrillow, 2015. 'The Forgotten Expedition—1885: The Strickland River, New Guinea.' *Journal of the Royal Australian Historical Society* 101: 7–24.

Dwyer, P.D., M. Minnegal and V. Woodyard, 1993. 'Konai, Febi and Kubo: The Northwest Corner of the Bosavi Language Family.' *Canberra Anthropology* 16: 1–14 (doi.org/10.1080/03149099309508439).

Englund, H. and J. Leach, 2000. 'Ethnography and the Meta-Narratives of Modernity.' *Current Anthropology* 41: 225–248.

Ernst, T.M., 1999. 'Land, Stories, and Resources: Discourse and Entification in Onabasulu Modernity.' *American Anthropologist* 101: 88–97 (doi.org/10.1525/aa.1999.101.1.88).

——, 2008. 'Full-Scale Social Mapping and Landowner Identification Study of PRL02.' Unpublished report to ExxonMobil Corporation.

Errington, F. and D. Gewertz, 1996. 'The Individuation of Tradition in a Papua New Guinean Modernity.' *American Anthropologist* 98: 114–126 (doi.org/10.1525/aa.1996.98.1.02a00100).

Everill, C.E., 1888. 'Exploration of New Guinea - Capt. Everill's report.' *Transactions and Proceedings of the Royal Geographical Society of Australasia, NSW Branch*: 170–186.

Filer, C., 1997. 'The Melanesian Way of Menacing the Mining Industry.' In B. Burt and C. Clerk (eds), *Environment and Development in the Pacific Islands*. Canberra: Australian National University, National Centre for Development Studies (Pacific Policy Paper 25).

——, 2011a. 'The New Land Grab in Papua New Guinea: A Case Study from New Ireland Province.' Canberra: The Australian National University, State Society and Governance in Melanesia Program (Discussion Paper 2011/2).

——, 2011b. 'The Political Construction of a Land Grab in Papua New Guinea.' Canberra: Australian National University, Crawford School of Economics and Government (READ Pacific Discussion Paper 1).

Fitzpatrick, P., 1971. 'Nomad Patrol Report No. 8 of 1970/71: Western District.' Territory of Papua and New Guinea.

——, 2010. 'Preliminary Social Mapping and Landowner Identification Study Petroleum Prospecting Licence 287: A Report for Sasol Petroleum Papua New Guinea Ltd.' Port Moresby: Firewall Logistics Ltd.

Fletcher, L. and A. Webb, 2012. *Pipe Dreams: The PNG LNG Project and the Future Hopes of a Nation*. Sydney: Jubilee Australia Research Centre.

Franklin, K.J., 1972. 'A Ritual Pandanus Language of Papua New Guinea.' *Oceania* 43: 66–76 (doi.org/10.1002/j.1834-4461.1972.tb01197.x).

Froggatt, W.W., 1936. 'New Guinea 50 Years Ago: Records from My Old Diary Kept During the Geographical Society of Australasia's Expedition to the Strickland River, New Guinea 1885.' Unpublished manuscript in University of Papua New Guinea Library (Special Collection AL-4).

Gabriel, J. and M. Wood, 2015. 'The Rimbunan Hijau Group in the Forests of Papua New Guinea.' *The Journal of Pacific History* 50: 322–343 (doi.org/10.1080/00223344.2015.1060925).

Gammage, B., 1998. *The Sky Travellers: Journeys in New Guinea 1938–1939.* Carlton: Melbourne University Press.

Garnaut, J., 2015. 'PNG Chiefs Talk of Civil Unrest over Unpopular Australian Bank Deal.' *Sydney Morning Herald Business Day,* 11 October.

Gell, A., 1998. *Art and Agency: An Anthropological Theory.* Oxford: Clarendon Press.

Gentleman, J., 2015. 'Meant to Keep Malaria Out, Mosquito Nets Are Used to Haul Fish In.' *New York Times,* 24 January.

Gerber, N. and R.J. Hill, 2005. 'Sustainability and Market Structure in Renewable Natural Resource Markets: The Case of Gaharu in Papua New Guinea.' Viewed 22 August 2014 at: een.anu.edu.au/e05prpap/gerber.pdf

Gewertz, D. and F. Errington, 2016, 'Retelling Chambri Lives: Ontological Bricolage.' *The Contemporary Pacific* 28: 347–381 (doi.org/10.1353/cp.2016.0031).

Gibson, J.J., 1979. *The Ecological Approach to Visual Perception.* Boston: Houghton Mifflin.

Gilberthorpe, E. and G. Banks, 2012. 'Development on Whose Terms?: CSR Discourse and Social Realities in Papua New Guinea's Extractive Industries Sector.' *Resources Policy* 37: 185–193 (doi.org/10.1016/j.resourpol.2011.09.005).

Goldman, L., 2007. '"Hoo-ha" in Huli: Considerations on Commotion and Community in the Southern Highlands.' In N. Haley and R.J. May (eds), *Conflict and Resource Development in the Southern Highlands of Papua New Guinea.* Canberra: ANU E Press.

Goldman, L. (ed.), 2009. *Papua New Guinea Liquefied Natural Gas Project: Social Impact Assessment 2008.* Report to ExxonMobil Corporation.

Goldman, L. and T. Ernst, 2008. 'Full-scale Social Mapping and Landowner Identification Study of Proposed PNG LNG Gas Project Pipeline Route PRLs 02–12.' Unpublished report to ExxonMobil Corporation.

Golub, A., 2006. 'Who Is the "Original Affluent Society"? Ipili "Predatory Expansion" and the Porgera Gold Mine, Papua New Guinea.' *The Contemporary Pacific* 18: 265–292 (doi.org/10.1353/cp.2006.0016).

———, 2007. 'Ironies of Organization: Landowners, Land Registration, and Papua New Guinea's Mining and Petroleum Industry.' *Human Organization* 66: 38–48 (doi.org/10.17730/humo.66.1.157563342241q348).

———, 2014. *Leviathans at the Gold Mine: Creating Indigenous and Corporate Actors in Papua New Guinea.* Durham (NC): Duke University Press (doi.org/10.1215/9780822377399).

GoPNG (Government of Papua New Guinea), 2005. 'National HIV/AIDS Support Project. Situational Analysis for Strategic Planning at District Level, Western Province.' National AIDS Council, Papua New Guinea.

———, 2009. 'PNG LNG Project: Umbrella Benefits Sharing Agreement 2009.' Viewed 18 November 2015 at: actnowpng.org/sites/default/files/UBSA.pdf

———, 2012a. 'Incorporated Land Groups Guidelines.' Port Moresby: Department of Lands and Physical Planning.

———, 2012b. *Training Manual 1 in Implementation of the Land Group Incorporation (Amendment) Act 2009 and the Land Registration (Amendment) Act 2009.* Port Moresby: Constitutional and Law Reform Commission.

———, 2014a. 'Business Names Act 2014.' Viewed 31 May 2016 at: www.ipa.gov.pg/wp-content/uploads/Bus-Names-Act-2014.pdf

———, 2014b. 'Companies (Amendment) Act 2014.' Viewed 31 May 2016 at: www.ipa.gov.pg/wp-content/uploads/ComNamesACT_-2014d.pdf

Gow, P., 1995. 'Land, People and Paper in Western Amazonia.' In E. Hirsch and M. O'Hanlon (eds), *The Anthropology of Landscape: Perspectives on Place and Space*. Oxford: Clarendon.

Guddemi, P., 1992. 'When Horticulturalists Are Like Hunter-gatherers: The Sawiyano of Papua New Guinea.' *Ethnology* 31: 303–314 (doi.org/10.2307/3773422).

Hage, G., 2012. 'Critical Anthropological Thought and the Radical Political Imaginary Today.' *Critique of Anthropology* 32(3): 285–308 (doi.org/10.1177/0308275X12449105).

Harrison, S., 1990. *Stealing People's Names: History and Politics in a Sepik River Cosomology*. Cambridge: Cambridge University Press (doi.org/10.1017/CBO9780511521096).

Hays, T., 1993. '"The New Guinea Highlands": Region, Culture Area, or Fuzzy Set.' *Current Anthropology* 34: 141–164 (doi.org/10.1086/204150).

Hides, J.G., 1939. *Beyond the Kubea*. Sydney: Angus and Robertson.

Hoad, R.A., 1964. 'Nomad Patrol Report 4/63-64: East Strickland Division, Western District.' Territory of Papua and New Guinea.

Hobsbawm, E.J. and T. Ranger (eds), 1983. *The Invention of Tradition*. Cambridge: Cambridge University Press.

Ingold, T., 2000. *The Perception of the Environment: Essays in Livelihood, Dwelling and Skill*. London: Routledge (doi.org/10.4324/9780203466025).

———, 2010. 'Bringing Things to Life: Creative Entanglements in a World of Materials.' Manchester, University of Manchester, ESRC National Centre for Research Methods (Realities Working Paper 15).

Jacka, J., 2015. *Alchemy in the Rain Forest: Politics, Ecology, and Resilience in a New Guinea Mining Area*. Durham (NC): Duke University Press (doi.org/10.1215/9780822375012).

Jackson, R.T., 2015. 'The Development and Current State of Landowner Businesses Associated with Resource Projects in Papua New Guinea.' Port Moresby: Papua New Guinea Chamber of Mines and Petroleum.

Johnson, A.E., 1968. 'Nomad Patrol Report 2/68-69: Cecilia and Carrington River Areas, Western District.' Territory of Papua and New Guinea.

Jolly, M. and N. Thomas, 1992. 'Introduction: The Politics of Tradition in the Pacific.' *Oceania* 62: 241–248 (doi.org/10.1002/j.1834-4461.1992.tb00355.x).

Jorgensen, D., 1993. 'Money and Marriage in Telefolmin: From Sister Exchange to Daughter as Trade Store.' In R.A. Marksbury (ed.), *The Business of Marriage: Transformations in Oceanic Matrimony.* Pittsburgh: University of Pittsburgh Press (ASAO Monograph 4).

——, 1997. 'Who or What Is a Landowner?: Mythology and Marking the Ground in a Papua New Guinea Mining Project.' *Anthropological Forum* 7: 599–627 (doi.org/10.1080/00664677.1997.9967476).

JTA International, 2009. 'North Fly Health Services Development Program: 2009 Annual Report.' Viewed 26 July 2015 at: www.nfhsdp. org/wp-content/uploads/2013/11/NFHSDP-2009-Annual-Report-with-Annexes.pdf

Kalinoe, L.K., 2003. 'Incorporated Land Groups in Papua New Guinea.' *Melanesian Law Journal.* Viewed 30 May 2016 at: www.paclii.org/journals/MLJ/2003/4.html

Karius, C., 1928. 'Appendix A: Report of North-West Patrol.' In *Territory of Papua Annual Report for the Year 1926–27*. Canberra: Government Printer.

Keane, W. 2013. 'Ontologies, anthropologists, and ethical life.' *HAU: Journal of Ethnographic Theory* 3: 186–91 (doi.org/10.14318/hau3.1.014).

Keesing, R., 1989. 'Creating the Past: Culture and Tradition in the Contemporary Pacific.' *The Contemporary Pacific* 1: 19–42.

Kirsch, S., 2006. *Reverse Anthropology: Indigenous Analysis of Social and Environmental Relations in New Guinea.* Palo Alto (CA): Stanford University Press.

——, 2014. *Mining Capitalism: The Relationship between Corporations and Their Critics.* Berkeley: University of California Press.

Klein, N., 2000. *No Logo, No Space, No Choice, No Jobs: Taking Aim at the Brand Bullies*. London: Flamingo.

Knauft, B.M., 1985. *Good Company and Violence: Sorcery and Social Action in a Lowland New Guinea Society*. Berkeley: University of California Press.

——, 1987. 'Managing Sex and Anger: Tobacco and Kava Use among the Gebusi of Papua New Guinea.' In L. Lindstrom (ed.), *Drugs in Western Pacific Societies: Relations of Substance*. New York: University Press of America (ASAO Monograph 11).

—— (ed.), 2002a. *Critically Modern: Alternatives, Alterities, Anthropologies*. Bloomington: Indiana University Press.

——, 2002b. *Exchanging the Past: A Rainforest World of Before and After*. Chicago: University of Chicago Press.

——, 2007. 'From Self-Decoration to Self-Fashioning: Orientalism as Backward Progress among the Gebusi of Papua New Guinea.' In E. Ewart and M. O'Hanlon (eds), *Body Arts and Modernity*. Wantage: Sean Kingston.

——, 2010. 'Gebusi Religion and Conversion Revisited: Spiritual Change in the Area of Nomad Station, Western Province, Papua New Guinea.' *Asia-Pacific Forum* 48: 8–30.

——, 2011. 'Men, Modernity and Melanesia.' In D. Lipset and P. Roscoe (eds), *Echoes of the Tambaran: Masculinity, History and the Subject in the Work of Donald F. Tuzin*. Canberra: ANU E Press.

——, 2016. *Gebusi: Lives Transformed in a Rainforest World* (4th edition). Longrove (IL): Waveland Press.

Koim, S. and S. Howes, 2016. 'PNG LNG Landowner Royalties – Why so Long?' Devpolicy blogpost, 16 December. Viewed 10 January 2016 at: devpolicy.org/png-lng-landowner-royalties-long-20161216/

Koyama, S.K., 2004, 'Reducing Agency Problems in Incorporated Land Groups.' *Pacific Economic Bulletin* 19: 20–31.

Kuchikura, Y., 1995. 'Productivity and adaptability of diversified food-getting system of a foothill community in Papua New Guinea.' *Bulletin of the Faculty of General Education, Gifu University* 31: 45–76.

Lang, M., 1961a. 'Interim Report – Strickland Patrol (5 October 1961).' Patrol Reports, Kiunga, Western District. Volume 10: 1961–1962. Territory of Papua and New Guinea.

——, 1961b. 'Interim Report – Strickland Patrol (17 October 1961).' Patrol Reports, Kiunga, Western District. Volume 10: 1961–1962. Territory of Papua and New Guinea.

——, 1961c. 'Strickland Patrol – Fourth Interim Report (23 May 1962).' Patrol Reports, Kiunga, Western District. Volume 10: 1961–1962. Territory of Papua and New Guinea.

——, 1962a. 'Strickland Patrol – Murder of Constable Kasika 9259 (9 November 1961).' Patrol Reports, Kiunga, Western District. Volume 10: 1961–1962. Territory of Papua and New Guinea.

——, 1962b. 'Special Patrol Report – Mr. Stott.' Patrol Reports, Kiunga, Western District. Volume 10: 1961–1962. Territory of Papua and New Guinea.

Latour, B., 1999. *Pandora's Hope: Essays on the Reality of Science Studies.* Cambridge (MA): Harvard University Press.

Lattas, A., 2011. 'Logging, Violence and Pleasure: Neoliberalism, Civil Society and Corporate Governance in West New Britain.' *Oceania* 81: 88–107 (doi.org/10.1002/j.1834-4461.2011.tb00095.x).

Lea, D., 2013. 'A Critical Review of the Recent Amendments to the Customary Land Tenure System in Papua New Guinea.' *Social Development Issues* 35(3): 63–81.

MacCarthy, M., 2013. '"More than Grass Skirts and Feathers": Negotiating Culture in the Trobriand Islands.' *International Journal of Heritage Studies* 19: 62–77 (doi.org/10.1080/13527258.2011.637 946).

Macdonald-Smith, A., 2015. 'Oil Search to Reap PNG LNG Cash Flows as Project Deemed Complete.' *Sydney Morning Herald Business Day*, 6 February.

Macintyre, M., 2007. 'Informed Consent and Mining Projects: A View from Papua New Guinea.' *Pacific Affairs* 80: 49–65 (doi.org/ 10.5509/200780149).

——, 2008. 'Police and Thieves, Gunmen and Drunks: Problems with Men and Problems with Society in Papua New Guinea.' *The Australian Journal of Anthropology* 19: 179–193 (doi.org/10.1111/j.1835-9310.2008.tb00121.x).

——, 2011. 'Modernity, Gender and Mining: Experiences from Papua New Guinea.' In K. Lahiri-Dutt (ed.), *Gendering the Field: Towards Sustainable Livelihoods for Mining Communities*. Canberra: ANU E Press.

Mackay, R.D., 2012. *The Bonito Expedition: The New Guinea Exploring Expedition of 1885*. Belair (SA): Crawford House Publishing.

Malafouris L. and C. Renfrew (eds), 2010. *The Cognitive Life of Things: Recasting the Boundaries of the Mind*. Cambridge: McDonald Institute for Archaeological Research Publications.

Martin, K., 2013. *The Death of the Big Men and the Rise of the Big Shots: Custom and Conflict in East New Britain*. Oxford: Berghahn Books.

Maurer, B., 2006. 'The Anthropology of Money.' *Annual Review of Anthropology* 35: 15–36 (doi.org/10.1146/annurev.anthro.35.081705.123127).

May, R.J. and M. Spriggs, 1990. *The Bougainville Crisis*. Bathurst (NSW): Crawford House Publishing.

McBride, B., 1960. 'Kiunga Patrol Report No. 2 of 59/60: Western District.' Territory of Papua and New Guinea.

McGee, W.A., 2007. 'Tragedy on the Strickland: Jack Hides and the Investors Ltd Expedition of 1937.' *Journal of Australasian Mining History* 5: 150–170.

McGregor, J.K., 1969. 'Kiunga Patrol Report No. 10 of 1968-1969: Western District.' Territory of Papua and New Guinea.

McIlriath, J., S. Robinson, L.L. Pyrambone, L. Petai, D. Sinebare and S. Maiap, 2012. 'The Community Good: Examining the Influence of the PNG LNG Project in the Hela Region of Papua New Guinea.' Dunedin: University of Otago, National Centre for Peace and Conflict Studies.

McPhail, M.L., 1991. 'Complicity: The Theory of Negative Difference.' *Howard Journal of Communications* 3: 1–13 (doi.org/10.1080/10646179109359734).

Meintjes, L.A., 1972. 'Nomad Patrol Report 24 of 1971/72, Western District.' Territory of Papua and New Guinea.

——, 1973. 'Nomad Patrol Report 1 of 1973/74, Western District.' Territory of Papua and New Guinea.

Mimica, J., 1988. *Intimations of Infinity: The Cultural Meanings of the Iqwaye Counting System and Number.* Oxford: Berg.

Minakawa, N., G.O. Dida, G.O. Sonye, K. Futami and S. Kaneko, 2008. 'Unforeseen Misuses of Bed Nets in Fishing Villages along Lake Victoria.' *Malaria Journal* 7: 165 (doi.org/10.1186/1475-2875-7-165).

Minnegal, M., 1994. Fishing at Gwaimasi: The Interaction of Social and Ecological Factors in influencing Subsistence Behaviour. St Lucia: University of Queensland (PhD thesis).

——, 1997. 'Consumption and Production: Sharing and the Social Construction of Use-Value.' *Current Anthropology* 38: 25–48 (doi.org/10.1086/204580).

——, 2009. 'The Time Is Right: Waiting, Reciprocity and Sociality.' In G. Hage (ed.), *Waiting.* Carlton: Melbourne University Press.

Minnegal, M. and P.D. Dwyer, 1997. 'Women, Pigs, God and Evolution: Social and Economic Change among Kubo People of Papua New Guinea.' *Oceania* 68: 47–60 (doi.org/10.1002/j.1834-4461.1997.tb02641.x).

——, 1998. 'Intensification and Complexity in the Interior Lowlands of Papua New Guinea: A Comparison of Bedamuni and Kubo.' *Journal of Anthropological Archaeology* 17: 375–400 (doi.org/10.1006/jaar.1998.0327).

——, 1999. 'Re-reading Relationships: Changing Constructions of Identity among Kubo of Papua New Guinea.' *Ethnology* 38: 59–80 (doi.org/10.2307/3774087).

——, 2000a. 'Responses to a Drought in the Tropical Lowlands of Papua New Guinea: A Comparison of Bedamuni and Kubo-Konai.' *Human Ecology* 28: 493–526 (doi.org/10.1023/A:1026483630039).

——, 2000b. 'A Sense of Community: Sedentary Nomads of the Interior Lowlands of Papua New Guinea.' *People and Culture in Oceania* 16: 43–65.

——, 2001. 'Intensification, Complexity and Evolution: Insights from the Strickland-Bosavi Region.' *Asia Pacific Viewpoint* 42: 269–285 (doi.org/10.1111/1467-8373.00149).

——, 2006. 'Fertility and Social Reproduction in the Strickland-Bosavi Region.' In S.J. Ulijaszek (ed.), *Population, Reproduction and Fertility in Melanesia*. Oxford: Berghahn Books.

——, 2007. 'Money, Meaning and Materialism: A Papua New Guinean Case History.' Melbourne: The University of Melbourne, School of Social & Environmental Enquiry (Working Paper in Development 2/2007).

——, 2011a. 'Boundaries and Barriers among Kubo and Beyond.' *Journal of the Polynesian Society* 120: 315–331.

——, 2011b. 'Appropriating Fish, Appropriating Fishermen: Tradable Permits, Natural Resources and Uncertainty.' In V. Strang and M. Busse (eds), *Appropriation and Ownership*. Oxford: Berg (ASA Monograph 47).

Minnegal, M., S. Lefort and P.D. Dwyer, 2015. 'Reshaping the Social: A Comparison of Fasu and Kubo-Febi Approaches to Incorporating Land Groups.' *The Asia Pacific Journal of Anthropology* 16: 496–513 (doi.org/10.1080/14442213.2015.1085078).

Mirou, N., 2013. *Commission of Inquiry into Special Agriculture and Business Lease (C.O.I. SABL): Report*. Port Moresby: Government of Papua New Guinea. Viewed 26 June 2014 at: www.coi.gov.pg/documents/COI%20 SABL/Mirou%20SABL%20Final%20Report.pdf

Moi, T., 1991. 'Appropriating Bourdieu: Feminist Theory and Pierre Bourdieu's Sociology of Culture.' *New Literary History* 22: 1017–1049 (doi.org/10.2307/469077).

Morphy, H., 2009. 'Art as a Mode of Action.' *Journal of Material Culture* 14: 5–27 (doi.org/10.1177/1359183508100006).

Mosko, M., 2010. 'Partible Penitents: Dividual Personhood and Christian Practice in Melanesia and the West.' *Journal of the Royal Anthropological Institute* (N.S.) 16: 215–240 (doi.org/10.1111/j.1467-9655.2010.01618.x).

Naveh, D. and N. Bird-David, 2014. 'How Persons Become Things: Economic and Epistemological Changes among Nayaka Hunter-Gatherers.' *Journal of the Royal Anthropological Institute* (N.S.) 20: 74–92 (doi.org/10.1111/1467-9655.12080).

Nelson, P.N., J. Gabriel, C. Filer, M. Banabas, J. A. Sayer, G. N. Curry, G. Koczberski and O. Venter, 2014. 'Oil Palm and Deforestation in Papua New Guinea.' *Conservation Letters* 7: 188–195 (doi.org/10.1111/conl.12058).

Norgan, N.G., 1995. 'Changes in Patterns of Growth and Nutritional Anthropometry in Two Rural Modernizing Papua New Guinea Communities.' *Annals of Human Biology* 22: 491–513 (doi.org/10.1080/03014469500004162).

Numapo, J., 2013. *Commission of Inquiry into the Special Agriculture and Business Lease (SABL): Final Report.* Port Moresby: Government of Papua New Guinea. Viewed 26 June 2014 at: www.coi.gov.pg/documents/COI%20SABL/Numapo%20SABL%20Final%20Report.pdf

Odani, S., 2002. 'Subsistence Ecology of the Slash and Mulch Cultivating Method: Empirical Study in the Great Papuan Plateau of Papua New Guinea.' *People and Culture in Oceania* 18: 45–63.

Oil Search Ltd., 2015. 'Annual Report 2015.' Viewed 31 May 2017 at: www.oilsearch.com/__data/assets/pdf_file/0018/1566/OSH_AR16-a3d8b18e-1b65-4a66-a4a0-34826fa5db87-2.PDF

——, 2016a. 'Exploration and Appraisal Drilling Update – June 2016.' Viewed 13 August 2016 at: www.oilsearch.com (see ASX Releases).

——, 2016b. 'Exploration and Appraisal Drilling Update – October 2016.' Viewed 5 November 2016 at: www.oilsearch.com (see ASX Releases).

——, 2016c. 'Exploration and Appraisal Drilling Update – November 2016.' Viewed 13 December 2016 at: www.oilsearch.com (see ASX Releases).

Ortner, S.B., 1995. 'Resistance and the Problem of Ethnographic Refusal.' *Comparative Studies in Society and History* 37: 173–193 (doi.org/10.1017/S0010417500019587).

Otto, T. and R.J. Verloop, 1996. 'The Asaro Mudmen: Local Property, Public Culture?' *The Contemporary Pacific* 8: 349–386.

Pálsson, G., 1994. 'Enskilment at Sea.' *Man* (N.S.) 29: 901–927 (doi.org/10.2307/3033974).

Patterson, W.R., 1969a. 'Nomad Patrol Report No. 19 of 1968–69: Western District.' Territory of Papua New Guinea.

——, 1969b. 'Nomad Patrol Report No. 14, 1968–69: Western District.' Territory of Papua New Guinea.

PNG (Papua New Guinea) Today, 2014. 'O'Neill: Economy to Pick Up, Critics Have No Answers.' Viewed 25 August 2014 at: news.pngfacts.com/2014/08/oneill-economy-to-pick-up-critics-have.html

Povinelli, E.A., 1995. 'Do Rocks Listen? The Cultural Politics of Apprehending Australian Aboriginal Labor.' *American Anthropologist* 97: 505–518 (doi.org/10.1525/aa.1995.97.3.02a00090).

——, 2001. 'Radical Worlds: The Anthropology of Incommensurability and Inconceivability.' *Annual Reviews of Anthropology* 30: 319–334 (doi.org/10.1146/annurev.anthro.30.1.319).

Ratner, C., 2000. 'Agency and Culture.' *Journal for the Theory of Social Behavior* 30: 413–434 (doi.org/10.1111/1468-5914.00138).

Roepstorff, A., 2003. 'Clashing Cosmologies: Contrasting Knowledge in the Greenlandic Fishery.' In A. Roepstorff, N. Bubandt and K. Kull (eds), *Imagining Nature: Practices of Cosmology and Identity*. Langelandgade: Aarhus University Press.

Rouzet, C., 2013. 'Exxon Mobil in Papua New Guinea: Shady Stories at the Holiday Inn.' Viewed 26 June 2014 at: pulitzercenter.org/reporting/papua-new-guinea-shady-stories-holiday-inn-exxon-mobil-rouzet

Russell, P.J., 1962. 'Kiunga Patrol Report No. 1 of 1961/62: Western District.' Territory of Papua and New Guinea.

Sakata, H. and B. Prideaux, 2013. 'An Alternative Approach to Community-Based Ecotourism: A Bottom-Up Locally Initiated Non-Monetised Project in Papua New Guinea.' *Journal of Sustainable Tourism* 21: 880–899 (doi.org/10.1080/09669582.2012.756493).

Salim, E., 2004. 'Striking a Better Balance – Volume 1: The World Bank Group and Extractive Industries.' Jakarta: Extractive Industries Review.

Salisbury, R.F., 1970. *Vunamami: Economic Transformation in a Traditional Society*. Carlton: Melbourne University Press.

Scott, C., 1996. 'Science for the West, Myth for the Rest? The Case of James Bay Cree Knowledge Construction.' In L. Nader (ed.), *Naked Science: Anthropological Enquiry into Boundaries, Power, and Knowledge*. New York: Routledge.

Seymour, S., 2006. 'Resistance.' *Anthropological Theory* 6: 303–21 (doi.org/10.1177/1463499606066890).

Sharma, D., 2014. 'LNG Project Absorbs Skilled Workers in PNG.' *Islands Business*, 20 June.

Shaw, R.D., 1990. *Kandila: Samo Ceremonialism and Interpersonal Relationships*. Ann Arbor: University of Michigan Press.

——, 1996. *From Longhouse to Village: Samo Social Change*. New York: Harcourt Brace College Publishers.

Silverman, E.K., 1999. 'Tourist Art as the Crafting of Identity in the Sepik River (Papua New Guinea).' In R.B. Phillips and C.B. Steiner (eds), *Unpacking Culture: Art and Commodity in Colonial and Postcolonial Worlds*. Berkeley: University of California Press.

Slife, B.D., 2004. 'Taking Practice Seriously: Toward a Relational Ontology.' *Journal of Theoretical and Philosophical Psychology* 24: 157–178 (doi.org/10.1037/h0091239).

Stanley, G.A.V., 1948. 'Report on the Fly-Strickland Survey, Permits 4 & 11, Papua (including an Appendix by M.F. Glaessner and P.J. Coleman titled 'Report on the Palaeontological Examination of Rock Samples from the Strickland Survey').' Sydney: Oil Search Library (Reference RP05474).

Stone, B., 2012. 'Connections.' Oil Search Ltd. Viewed 31 May 2017 at: www.oilsearch.com/__data/assets/pdf_file/0014/6044/OSH-CASE-STUDY-BOOK.pdf

Stott, R.R., 1962. 'Kiunga Patrol Reports 1961-62: Special Report.' [Patrol Reports, Kiunga, Western District. Volume 10: 1961–1962.] Territory of Papua and New Guinea.

Strathern, A. and P.J. Stewart, 2004. *Empowering the Past, Confronting the Future: The Duna People of Papua New Guinea*. New York: Palgrave Macmillan (doi.org/10.1057/9781403982421).

Strathern, M., 1988. *The Gender of the Gift: Problems with Women and Problems with Society in Melanesia*. Berkeley: University of California Press (doi.org/10.1525/california/9780520064232.001.0001).

Suda, K., 1990. 'Leveling Mechanisms in a Recently Relocated Kubor Village, Papua New Guinea: A Socio-Behavioral Analysis of Sago-Making.' *Man and Culture in Oceania* 6: 99–112.

Taprin, H., 2007/8. 'The Origin of Kesomo.' Unpublished manuscript.

Taylor, A.-C., 2013. 'Distinguishing Ontologies.' *HAU: Journal of Ethnographic Theory* 3: 201–204 (doi.org/10.14318/hau3.1.017).

Temu, P. and P.C.Y. Chen, 1999. 'The Papua New Guinea Vision of Healthy Islands.' *Pacific Health Dialog* 6: 253–258.

Thompson, C., 2011. 'Final Frontier: Newly Discovered Species of New Guinea (1998–2008).' WWF Western Melanesia Programme Office. Viewed 14 October 2015 at d2ouvy59p0dg6k.cloudfront.net/downloads/new_guinea_new_species_2011.pdf

Tsing, A., 2004. *Friction: An Ethnography of Global Connection*. Princeton (NJ): Princeton University Press.

——, 2009. 'Supply Chains and the Human Condition.' *Rethinking Marxism* 21: 148–176 (doi.org/10.1080/08935690902743088).

——, 2013. 'Sorting Out Commodities: How Capitalist Value is Made through Gifts.' *HAU: Journal of Ethnographic Theory* 3: 21–43 (doi.org/10.14318/hau3.1.003).

Ulijaszek, S.J., 1993. 'Evidence for a Secular Trend in Heights and Weights of Adults in Papua New Guinea.' *Annals of Human Biology* 20: 349–355 (doi.org/10.1080/03014469300002752).

——, 2003. 'Socio-Economic Factors Associated with Physique of Adults of the Purari Delta of the Gulf Province, Papua New Guinea.' *Annals of Human Biology* 30: 316–328 (doi.org/10.1080/03014460310000 86004).

Vigh, H., 2009. 'Motion Squared: A Second Look at the Concept of Social Navigation.' *Anthropological Theory* 9: 419–438 (doi.org/ 10.1177/1463499609356044).

Viveiros de Castro, E.B., 1998. 'Cosmological Deixis and Amerindian Perspectivism.' *Journal of the Royal Anthropological Institute* (N.S.) 4: 469–488 (doi.org/10.2307/3034157).

——, 2004a. 'Exchanging Perspectives: The Transformation of Objects into Subjects in Amerindian Ontologies.' *Common Knowledge* 10: 463–484 (doi.org/10.1215/0961754X-10-3-463).

——, 2004b. 'Perspectival Anthropology and the Method of Controlled Equivocation.' *Tipití: Journal of the Society for the Anthropology of Lowland South America* 2: 3–22.

——, 2012. 'Cosmological Perspectivism in Amazonia and Elsewhere.' *HAU Masterclass Series* 1. Viewed 26 March 2016 at: haubooks.org/ cosmological-perspectivism-in-amazonia/

Wagner, R., 1977. 'Scientific and Indigenous Papuan Conceptualizations of the Innate: A Semiotic Critique of the Ecological Perspective.' In T. Bayliss-Smith and R. Feachem (eds), *Subsistence and Survival: Rural Ecology in the Pacific*. London: Academic Press (doi.org/10.1016/ B978-0-12-083250-7.50018-2).

Walker, H., 2013. *Under a Watchful Eye: Self, Power, and Intimacy in Amazonia*. Berkeley: University of California Press.

Wardlow, H., 2004. 'The Mount Kare Python: Huli Myths and Gendered Fantasies of Agency.' In A. Rumsey and J. Weiner (eds), *Mining and Indigenous Lifeworlds in Australia and Papua New Guinea*. Oxford, UK: Sean Kingston Publishing.

Warrillow, C., 2007. 'The Future of Resource Development in the Southern Highlands.' In N. Haley and R.J. May (eds), *Conflict and Resource Development in the Southern Highlands of Papua New Guinea*. Canberra: ANU E Press.

Weiner, J.F., 2007. 'The Foi Incorporated Land Group: Group Definition and Collective Action in the Kutubu Oil Project Area, Papua New Guinea.' In J.F. Weiner and K. Glaskin (eds), *Customary Land Tenure and Registration in Australia and Papua New Guinea: Anthropological Perspectives*. Canberra: ANU E Press.

——, 2013. 'The Incorporated What Group: Ethnographic, Economic and Ideological Perspectives on Customary Land Ownership in Contemporary Papua New Guinea.' *Anthropological Forum* 23: 94–106 (doi.org/10.1080/00664677.2012.736858).

Welker, M., 2016. 'No Ethnographic Playground: Mining Projects and Anthropological Politics.' *Comparative Studies in Society and History* 58: 577–586 (doi.org/10.1017/S0010417516000189).

West, P., 2006. *Conservation Is Our Government Now: The Politics of Ecology in Papua New Guinea*. Durham (NC): Duke University Press (doi.org/10.1215/9780822388067).

Whiteman, G. and K. Mamen, 2002. *Meaningful Consultation and Participation in the Mining Sector: A Review of the Consultation and Participation of Indigenous Peoples within the International Mining Sector*. Ottawa: The North-South Institute.

Wildman, W.J., 2010. 'An Introduction to Relational Ontology.' In J. Polkinghorne (ed.), *The Trinity and an Entangled World: Relationality in Physical Science and Theology*. Cambridge: Wm. B. Eerdmans Publishing Co.

Zeitlyn, D. and R. Just, 2014. *Excursions in Realist Anthropology: A Merological Approach*. Newcastle upon Tyne: Cambridge Scholars Publishing.

Zich, F. and J. Compton, 2002. 'Agarwood (Gaharu) Harvest and Trade in Papua New Guinea: A Preliminary Assessment.' TRAFFIC Oceania. Viewed 22 August 2014 at: dev.cites.org/sites/default/files/eng/com/pc/11/X-PC11-Inf._11.pdf

www.ingramcontent.com/pod-product-compliance
Lightning Source LLC
Chambersburg PA
CBHW050807270326
41926CB00026B/4598